MIX
Papier aus verantwortungsvollen Quellen
Paper from responsible sources
FSC® C105338

Viktorija Širvinskytė

Developing the Solar Energy Sector in Lithuania

Solar Energy Sector Development Strategy for Lithuania based on the experience of the European Union

Anchor Academic Publishing

Širvinskytė, Viktorija: Developing the Solar Energy Sector in Lithuania: Solar Energy Sector Development Strategy for Lithuania based on the experience of the European Union, Hamburg, Anchor Academic Publishing 2014

Buch-ISBN: 978-3-95489-330-0
PDF-eBook-ISBN: 978-3-95489-830-5
Druck/Herstellung: Anchor Academic Publishing, Hamburg, 2014

Coverbild: pixabay.com

Bibliografische Information der Deutschen Nationalbibliothek:
Die Deutsche Nationalbibliothek verzeichnet diese Publikation in der Deutschen Nationalbibliografie; detaillierte bibliografische Daten sind im Internet über http://dnb.d-nb.de abrufbar.

Bibliographical Information of the German National Library:
The German National Library lists this publication in the German National Bibliography. Detailed bibliographic data can be found at: http://dnb.d-nb.de

All rights reserved. This publication may not be reproduced, stored in a retrieval system or transmitted, in any form or by any means, electronic, mechanical, photocopying, recording or otherwise, without the prior permission of the publishers.

Das Werk einschließlich aller seiner Teile ist urheberrechtlich geschützt. Jede Verwertung außerhalb der Grenzen des Urheberrechtsgesetzes ist ohne Zustimmung des Verlages unzulässig und strafbar. Dies gilt insbesondere für Vervielfältigungen, Übersetzungen, Mikroverfilmungen und die Einspeicherung und Bearbeitung in elektronischen Systemen.

Die Wiedergabe von Gebrauchsnamen, Handelsnamen, Warenbezeichnungen usw. in diesem Werk berechtigt auch ohne besondere Kennzeichnung nicht zu der Annahme, dass solche Namen im Sinne der Warenzeichen- und Markenschutz-Gesetzgebung als frei zu betrachten wären und daher von jedermann benutzt werden dürften.

Die Informationen in diesem Werk wurden mit Sorgfalt erarbeitet. Dennoch können Fehler nicht vollständig ausgeschlossen werden und die Diplomica Verlag GmbH, die Autoren oder Übersetzer übernehmen keine juristische Verantwortung oder irgendeine Haftung für evtl. verbliebene fehlerhafte Angaben und deren Folgen.

Alle Rechte vorbehalten

© Anchor Academic Publishing, Imprint der Diplomica Verlag GmbH
Hermannstal 119k, 22119 Hamburg

Acknowledgement

Foremost, I have to say that, apart from the efforts of my own, the encouragement, guidelines and support of some other people were essential for the successful achievement of my goal - conducting a study on a really interesting to me topic, where I can analyse relative issues and find the answers to my questions. Due to this, I take this opportunity to express my gratitude to all of those, who contributed to my motivation.

I would like to thank everyone, who made it possible for me to become a double degree student: I am very grateful to Vilnius Gediminas technical university for the opportunity to study abroad and to Fachhochschule Stralsund for all the experience and knowledge I gained there. The education, experience and opportunities given to me by these two institutions led me to writing this book.

A special gratitude I want to express to Prof. Dr. phil. Dr. hc. Hiltgunt Fanning and Skirmantas Šidlauskas for the understanding and the help throughout the process of my research.

I also want to thank my friends, family and the loved ones for the support and encouragement. Without that, it would have been much harder to achieve my goal. Especially I am grateful to my mother for always encouraging me and to Michael Korotkov for the support, understanding and being there for me.

Additional thanks to Stefanie Wenzel, Anne Quander, Dr. Thomas Hausmann, Dr. Doc. Algita Miečinskienė, Dr. Indrė Lapinskaitė and Gražina Jurgilienė for all the help and support during my studies.

Table of Contents

List of Tables 8

List of Figures 8

List of Abbreviations 9

INTRODUCTION 10

1. THE RELEVANCE OF THE SOLAR ENERGY SECTOR.................................... 12

    1.1. Securing the energy supply 15

    1.2. Effect to the Economy 16

    1.3. Protecting the environment 16

    1.4. Solar energy technologies 17

    1.5. The energy production... 17

    1.6. The use of the energy 18

    1.7. Future solar energy demand 18

    1.8. The overview of the Solar Energy sectors of the EU countries 20

    1.9. Screening out the countries with the most advanced solar energy sector 21

    1.10. The short overview of chosen countries and their solar energy sectors.......... 22

        1.10.1. Germany.. 23

        1.10.2. The United Kingdom 24

        1.10.3. Italy 25

    1.11. Choosing the country for further examples and comparisons......................... 26

2. THE OVERVIEW OF THE CHOSEN COUNTRIES 29

    2.1. Case: Lithuania.. 29

        2.1.1. Current situation... 33

        2.1.2. Prospects and forecasts 35

        2.1.3. Current conditions for Business... 37

    2.2. Case: Germany 38

        2.2.1. Current situation... 40

 2.2.2. Prospects and forecasts 43

 2.2.3. Current conditions for Business 44

 2.3. Comparison of the analysed countries... 45

3. DEVELOPING THE MODEL FOR THE PROMOTION OF THE DEVELOPMENT OF THE SOLAR ENERGY SECTOR OF LITHUANIA 49

 3.1. The description of the model.. 49

  3.1.1. Adopting the Long-Term Strategy.. 50

  3.1.2. Improving the Feed-In Tariff System 51

  3.1.3. Improving the System of Limitation... 53

  3.1.4. Avoiding Making Sudden Changes 54

 3.2. Evaluating the model.. 54

CONCLUSIONS AND SUGGESTIONS... 55

BIBLIOGRAPHY 57

List of Tables

Table 1. PV tariffs for 2012 in Germany 24
Table 2. PV tariffs for 2011 and 2012 in the United Kingdom 25
Table 3. PV tariffs for 2012 in Italy 26
Table 4. PV tariffs for 2011 and 2012 in Lithuania 32
Table 5. PV tariffs for 3rd and 2nd quarter of 2013 in Lithuania 34
Table 6. Assessment of RES-E policies and measures: Lithuania vs. Germany 45
Table 7. PV tariffs of Lithuania and Germany 48

List of Figures

Fig 1. Estimate of renewable energy growth for the EU, 2006-2030, GWh 19
Fig 2. Evolution of European new grid-connected PV capacities 22
Fig 3. Renewable energy country attractiveness index scores and rankings at May 2013 27
Fig 4. Technology-specific indices and ranking of the countries 28
Fig 5. Direct payments to producers of electricity from renewable energy sources for electricity supplied to the grid in years 2005-2012 in millions of Litas (1 Litas = 0.289626196 Euros) 31
Fig 6. Forecast of power generation from renewable energies in Lithuania 36
Fig 7. Ease of doing business in Lithuania 2013 (ranks out of 185 economies) 37
Fig 8. The impact of EEG revision on PV electricity production (in GWh) 40
Fig 9. PV on the path to becoming a key pillar of a sustainable energy supply in Germany 41
Fig 10. CO_2 Savings through PV systems 42
Fig 11. Most recent PV tariffs in Germany 42
Fig 12. Ease of doing business in Germany 2013 (ranks out of 185 economies) 44
Fig 13. Lithuania and comparator economies ranked by ease of doing business 46
Fig 14. Lithuania and comparator economies ranked by ease of starting a business 47

List of Abbreviations

ct - cent

EEG - Erneuerbare Energien Gesetz (eng. *Renewable Energy Law*)

Etc. - et cetera (and others)

EU - European Union

GDP - Gross domestic product

GW - Gigawatt

GWh - Gigawatt-hour

Ibid - ibidem (the same place)

kW - Kilowatt

kWh - Kilowatt-hour

Lt - Litas

MW - Megawatt

MWh - Megawatt-hour

NREAP - National Renewable Energy Action Plan

PV - Photovoltaic(s)

UK - United Kingdom

US - United States

US$ - United States Dollar(s)

W - Watt(s)

INTRODUCTION

Due to the renewable energies nowadays being of a great relevance in the European Union and recent solar energy sector issues in Lithuania and its relevance to investors and the country, the aim of this study is to create a model that would suggest the measures, actions, improvements for the solar energy sector of Lithuania in order to promote a better development of the sector and achieve more effective results, improve the current situation and future prospects of the solar energy sector in Lithuania and increase the attractiveness of investing in it.

The model is planned to be based on existing examples of one of the countries of the European Union or on a model created by combining the best features of several models that are proven to be successful in one or several countries of the European Union adjusting it to be eligible for solar energy sector of Lithuania. In order to create the model in this study it is foreseen to make an overview of several European Union countries analysing the more relevant to this topic aspects and futures, while intending to compare the information and come to the conclusions, which of the countries has a better or the best system of the solar energy sector, that could be used as an example for the development model for this sector in Lithuania.

This study consists of the three main parts:

1) the overview of the literature on the relevance of the topic covering the overview of the most advanced solar energy sectors of the European Union as well as the importance of the field in the region and in general;

2) the analysis of the important and the most relevant to the topic aspects in the research covering the overview of Lithuania and Germany;

3) the practical part with a description of a created model for the better development of the solar energy sector of Lithuania that consists of the main ideas and suggestions for Lithuania how to increase the attractiveness of the sector for investors.

The aim of the study: analysis of the solar energy sector of Lithuania with the goal of screening out the main issues and finding the aspects with the need of improvement as well as conducting the research in order to find the best example of successfully functioning solar energy sector in the European Union to have a reasoned basis for following suggestions to improve solar energy sector development in Lithuania.

Tasks of the study:

- Make an overview and analysis of the scientific literature and other sources relevant to the topic of the study and structure the most relevant and important findings;
- Make a research in order to screen out the most advanced solar energy sector in the European Union to use for following examples;
- Conduct an analysis of the solar energy sector of Lithuania in order to screen out the main issues and get a better understanding of the current situation as well as importance and consequences of the recent changes;
- Compare the findings and results of the research and analysis of the chosen countries and screen out the most important aspects that have the biggest influence on the development of the solar energy sector and need an improvement for the better functioning of the sector in future and increase of the solar energy sector attractiveness for investors;
- Create a solar energy sector development model for Lithuania consisting of the reasonable suggestions for improvement of the most important aspects in order to increase the attractiveness of the sector for investors as well as the development of the sector in general;
- Draw the general conclusions of the analysis, research and the model and come up with general suggestions concerning the aim and tasks of this study.

Methods of research and analysis:

- Analysis of the most relevant scientific and other literature and sources;
- Analysis of the most relevant statistical data;
- Comparative analysis;
- Graphical representation of the relevant information;
- Practical calculations, comparisons and evaluations.

Practical use of the study: the research and analysis conducted in the study gives a better understanding of the relevance of the solar energy and the current situation as well as the importance of the recent changes in the sector in Lithuania. It allows screening out the main aspects of the sector that require improvement and, while comparing with good examples, come up with ideas for needed actions and present reasonable suggestions for the improvement of the sector development and the ways of increasing the investment attractiveness of the sector. Using the suggested method might help improving the current situation and create a better future for the sector.

1. THE RELEVANCE OF THE SOLAR ENERGY SECTOR

The Solar Energy sector is a part of the Renewable Energy sector. Therefore, to explain the importance and the relevance of the solar energy sector in the chosen countries of the European Union, it is essential to define the relevance of the Renewable Energy sector in this region.

Renewable Energy in the European Union is a very relevant subject. It is a high European Union priority to promote the electricity from renewable energy sources. This is based on several reasons such as aim of diversification and security of the energy supply as well as the goal to protect the environment. It is also an important factor in the influence to the social and economic cohesion.[1]

The European Union aims to increase the share of electricity produced from the renewable energy sources. The main goal is to reach the amount of at least 20% of the final energy consumption provided by renewable energy sources by the year 2020.[2] As a part of the EU climate strategy, this aim is highly relevant while complying with the commitments on reducing the greenhouse gas emissions.[3]

Starting from the White Paper on the renewable energy sources in 1997, where the EU set the target of increasing the energy consumption from the renewable energy sources to 12% by the year 2010, through the years the goal of the EU grew and became 21%, that is set in the "Directive of the European Parliament and of the Council of 27 September 2001 on the promotion of electricity from renewable energy sources in the internal electricity market" (Directive 2001/77/EC).[4]

The general aims of the Directive were the promotion of an increase of the contribution of renewable energy sources to electricity production in the domestic market for electricity as well as the creation of the basis for a future Community foundation in this regard.[5] It concerns electricity produced from non-fossil renewable energy sources for instance geothermal, tidal, wave, hydroelectric, sewage treatment gas, landfill gas, biomass, biogas, wind and solar energies.[6]

---

[1] **Europa.eu** *"Renewable energy: the promotion of electricity from renewable energy sources"*, 20.01.2011, viewed on 20.06.2013, available online at URL: http://europa.eu/legislation_summaries/energy/renewable_energy/l27035_en.htm
[2] **Lehmann Paul / Creutzig Felix / Ehlers Melf-Hinrich / Friedrichsen Nele / Heuson Clemens / Hirth Lion / Pietzcker Robert** "Carbon Lock-Out: Advancing Renewable Energy Policy in Europe", 15.02.2012, p.324
[3] **Europa.eu** *"Renewable energy: the promotion of electricity from renewable energy sources"*, 20.01.2011, viewed on 20.06.2013, available online at URL: http://europa.eu/legislation_summaries/energy/renewable_energy/l27035_en.htm
[4] **Europa.eu** *"Renewable energy: the promotion of electricity from renewable energy sources"*, 20.01.2011, viewed on 20.06.2013, available online at URL: http://europa.eu/legislation_summaries/energy/renewable_energy/l27035_en.htm
[5] **DIRECTIVE 2001/77/EC OF THE EUROPEAN PARLIAMENT AND OF THE COUNCIL of 27 September 2001 on the promotion of electricity produced from renewable energy sources in the internal electricity market**, *Article 1*, 2001, p.5, viewed on 23.06.2013, available online at URL: http://europa.eu/legislation_summaries/energy/renewable_energy/l27035_en.htm
[6] **Europa.eu** *"Renewable energy: the promotion of electricity from renewable energy sources"*, 20.01.2011, viewed on 20.06.2013, available online at URL:http://europa.eu/legislation_summaries/energy/renewable_energy/l27035_en.htm

From January 1st of the year 2012 the Directive 2001/77/EC is repealed by "DIRECTIVE 2009/28/EC OF THE EUROPEAN PARLIAMENT AND OF THE COUNCIL of 23 April 2009 on the promotion of the use of energy from renewable sources and amending and subsequently repealing Directives 2001/77/EC and 2003/30/EC" (Directive 2009/28/EC), which set new, more relevant, aims and conditions, improving the ones set in the earlier Directive and the Directive that was created in between of the mentioned two.[7]

Directive 2009/28/EC sets a common structure for the promotion and the production of electricity from renewable energy sources. It is a part of the energy and climate legislation package that contributes a legislative foundation for Community goals for greenhouse gas emission savings. The Directive promotes energy efficiency, the improvement of energy supply, energy consumption from renewable sources and the economic stimulation of a dynamic sector.[8]

Directive 2009/28/EC establishes mandatory national goals for the proportion of energy from renewable sources in transport and for the overall share of energy from renewable sources in gross final consumption of energy. It sets the rules for: the statistical transfers between European Union countries; joint projects between its member states and with third countries; information and training; administrative procedures; guarantees of origin; access to the electricity grid for energy from renewable sources. The Directive also sets the sustainability criteria for bioliquids and biofuels.[9]

The Member States of the European Union are obliged to apply the provisions of the Directive on producing electricity from renewable energy sources. Each country sets the national indicative goals for the share of electricity produced from renewable energy sources the result of which has to meet the required amount. The Member States must, once in the established by the rules time period, publish reports that set the indicative targets of that particular country for future consumption of the electricity from renewable energy sources for the following ten years. The reports must also reveal the measures that are planned for meeting the targets or the measures used for the results achieved. The targets that every Member State of the EU sets must take account of the reference values set out in the documents attached to the Directive.[10]

---

[7] **Europa.eu** *"Renewable energy: the promotion of electricity from renewable energy sources"*, 20.01.2011, viewed on 20.06.2013, available online at URL: http://europa.eu/legislation_summaries/energy/renewable_energy/l27035_en.htm
[8] **Europa.eu** *"Renewable energy: Promotion of the use of energy from renewable sources"*, 09.07.2010, viewed on 22.06.2013, available online at URL: http://europa.eu/legislation_summaries/energy/renewable_energy/en0009_en.htm
[9] **DIRECTIVE 2009/28/EC OF THE EUROPEAN PARLIAMENT AND OF THE COUNCIL of 23 April 2009 on the promotion of the use of energy from renewable sources and amending and subsequently repealing Directives 2001/77/EC and 2003/30/EC**, *Article 1*, 2009, p.11, viewed on 23.06.2013, available online at URL: http://eur-lex.europa.eu/LexUriServ/LexUriServ.do?uri=OJ:L:2009:140:0016:0062:EN:PDF
[10] **Europa.eu** *"Renewable energy: the promotion of electricity from renewable energy sources"*, 20.01.2011, viewed on 20.06.2013, available online at URL: http://europa.eu/legislation_summaries/energy/renewable_energy/l27035_en.htm

The Directive aims to boost the contribution of the renewable energies when respecting the principles of the European market. Meanwhile, among the world leaders in development of new technologies connected with electricity from renewable energy sources are European companies.[11]

Concerning the relevance of the renewable energy and its sources it is explainable by its ability to fulfil all our energy needs, such as, producing electricity, heating houses and running transport. Based on the type of the renewable energy it can be used in different ways, for example, wind and hydro types of renewable energy are only used for generating electricity, when geothermal, biomass and solar energy is used for producing electricity as well as heat.[12] At this point, biomass, geothermal and solar energy can be defined and selected out from other renewable energies as the ones, that are more useful, therefore, more relevant and important.

During the year 2011 the two dominant renewable energy sectors were wind and solar (photovoltaic). Both of these sectors had record levels of installations: 42GW of wind and 25GW of solar. In these technologies alone was deployed more than US$100 billion of capital across a growing number of countries. In the year 2012 solar photovoltaics jumped to the top spot.[13]

As a conclusion from what was mentioned above, solar energy appears to be the most relevant part of the renewable energy sector in the EU from 2011-2012 years. Moreover, according to the data in 2013, solar energy reached more than 100GW of installed capacity through the year 2012. It is more than twice as much as it was two years ago.[14] This shows the growing interest in the solar energy sector resulting to its relevance and importance in the EU in general as well as in Lithuania and the other countries that were chosen for this study.

After establishing the relevance of the renewable energies in the European Union and its future strategies and singling out the energies that are used for a bigger range of purposes and suitable for production of more energy products than others, and after looking into the dominant energy sectors and recent European market tendencies we came to the conclusions of solar energy sector being the most relevant at the analysed period of time. Solar energy (or in other terms - photovoltaics) became a significant part of the EU electricity market while producing 2% of the demand and around 4% of peak demand in the region. Photovoltaics, for example, in Italy reached 5% of the electricity demand and over 10% of peak demand. In Germanys' southern federal state Bavaria the

---

[11] **Europa.eu** *"Renewable energy: the promotion of electricity from renewable energy sources"*, 20.01.2011, viewed on 20.06.2013, available online at URL:http://europa.eu/legislation_summaries/energy/renewable_energy/l27035_en.htm
[12] **European Commission**, *"Renewables make the difference"*, Luxembourg: Publications Office of the European Union, 2011, p.6, viewed on 23.06.2013, available online at URL:
http://ec.europa.eu/energy/publications/doc/2011_renewable_difference_en.pdf
[13] **KPMG International Cooperative**, *Green power 2012: The KPMG renewable energy M&A report*, 2012, p.13, p.29, viewed on 25.06.2013, available online at URL:
http://www.kpmg.com/CZ/cs/IssuesAndInsights/ArticlesPublications/Press-releases/Documents/KPMG-Green-Power-2012.pdf
[14] **Damian Carrington**, *Wind and solar power capacity surge*, 14 February 2013, viewed on 29.06.2013, available online at URL: http://www.euractiv.com/energy/wind-power-capacity-grew-20-glob-news-517720

capacity of the solar energy installations resulted to 600W per habitant, which is an astonishing amount. As well as for the other energy sources in the past, for photovoltaics to reach such level of development policy support was a crucial help.[15] In this chapter it is foreseen to analyse and describe this sector and its importance more thoroughly.

Firstly, it is important to reveal the reasons of such relevance of the solar energy, in order to understand its growing demand in the EU and the rest of the World. Some of the main reasons come from the features of solar energy that allow the absence of the noise, polluting gases or harmful emissions in the production. Furthermore, solar energy creates the possibility for the diversification of the energy supply and generates heat as well as electricity. Photovoltaics give an opportunity to create local jobs while stimulating the economy and the development of new technologies. Also one of the great features of the solar energy is its inexhaustibility and the fact that it is free. And last but not least, maintenance required for solar energy is minimal.[16]

Continuing the list of the benefits coming from the solar energy it is important that photovoltaic technologies are small, highly modular and are suitable for exploitation in any chosen location, when most of other electricity generation technologies have certain limitations. Furthermore, solar power is one of the renewable resources that are available all around the world and photovoltaics coincident with peak electricity demand that raises from cooling needs all year round in countries with common high temperatures and seasonal cooling (in summer) demand in other countries. In addition to that, photovoltaics have no fuel costs and their operation and maintenance costs are generally low. This is a big advantage while comparing it with conventional power plants, because it allows photovoltaics to offer competitive prices.[17]

1.1. Securing the energy supply

One of the most important reasons for the promotion of the development of solar energy sector in the EU is the need to secure the energy supply. The issue of the energy supply is highly important in this region based on its increasing dependency on fossil fuels (gas and oil) imports that are needed for the transport and electricity generation. Moreover, the EU relies on energy imports for almost half of its energy consumption. Based on the data collected in the 2011, fossil fuels stood for 79% of the regions' energy consumption. Ergo, the European Union can really benefit from the

---

[15] **European Photovoltaic Industry Association,** *Global Market Outlook For Photovoltaics Until 2016,* 2012, viewed on 10.06.2013, available online at URL:
http://www.helapco.gr/ims/file/reports/Global%20Market%20Outlook%202016.pdf
[16] **European Commission,** *Renewables Make the Difference,* 2011, p.15, viewed on 23.06.2013, available online at URL: http://www.energy.eu/publications/Renewables-make-the-difference-2011.pdf
[17] **International Renewable Energy Agency,** *Renewable Energy Technologies: Cost Analysis Series. Solar Photovoltaics,* 2012, viewed on 15.06.2013, available online at URL:
http://www.irena.org/DocumentDownloads/Publications/RE_Technologies_Cost_Analysis-SOLAR_PV.pdf

new ways of energy production, such as electricity production from renewable energy sources, and a raising number of energy suppliers in the region. Solar energy sector, as well as other renewable energy sectors, allows diversifying the energy supply in the EU, this way reducing the risks of supply cuts, price volatility and stimulating efficiency while increasing the competitiveness in the energy sector. According to the European Commission, renewable energy share amounting to 20% could reduce the fossil fuel imports of the EU by nearly 200 million tonnes of oil equivalent per year.[18]

1.2. Effect to the Economy

Another important reason for the solar energy sector development promotion in the EU is its effect to the economy. Solar energy, as well as other renewable energies, has a great potential for boosting the competitiveness of the European industry. It is highly important for the European Union to develop new low-carbon energy sources in order to avoid the substantial pollution and climate change costs. According to the European Commission, it is crucial for the economy of the EU to keep Europe between the leaders of such developments. Scientific know-how, technologies and development of renewable energy industry creates new value added jobs and strengthens the industry of the EU making it more competitive on a global scale. As mentioned earlier in this study, European companies dominate in the global renewable energy sector between manufacturers. According to the data of 2011, these companies employ more than 1.5 million people and have a turnover of more than 50 billion euro. If the strong growth of this sector continues, by the year 2020 it could provide an additional million jobs and might triple or at least double the mentioned turnover.[19]

1.3. Protecting the environment

Additional reason for the promotion of solar energy sector, as well as other renewable energy sectors, is the aim to protect the environment. The EU has set the target to reduce the pollution and is putting plenty of effort to control the issues of climate change. Current energy supply of the EU mainly consists of fossil fuels that give off greenhouse gases in the process of energy production. Concerning this issue, solar energy emits no gases of such sort in the process of the production, which makes such energy "friendly to the environment". Ergo, increasing the share of solar and other renewable energies in the total energy mix will result to reduction of greenhouse gas

---

[18] **European Commission**, *Renewables Make the Difference,* 2011, p.5, viewed on 23.06.2013, available online at URL: http://www.energy.eu/publications/Renewables-make-the-difference-2011.pdf

[19] **European Commission**, *Renewables Make the Difference,* 2011, p.4-5, viewed on 23.06.2013, available online at URL: http://www.energy.eu/publications/Renewables-make-the-difference-2011.pdf

emissions and help in the process of the environmental protection. As an addition to that, solar energy also helps to reduce air pollution, which has a direct influence on our health.[20]

1.4. Solar energy technologies

While analysing the solar energy sector, its relevance, advantages and other characteristics, one of the essential aspects appear to be the solar energy technologies. As mentioned earlier, solar energy technologies are also called photovoltaics. Photovoltaics (or solar cells) are electronic devices that produce electricity by converting sunlight directly into electric charge. In 1954 the modern form of such technologies was invented in Bell Telephone Laboratories. In the future global electricity generation mix the photovoltaic technologies are expected to play a major role based on it being the fastest growing renewable energy technologies in the present. The modular size of photovoltaic systems allows these technologies to be within the reach of small businesses, co-operatives and also individuals who seek for access to their own electricity generation and fixed electricity prices.[21]

1.5. The energy production

Using the solar energy technologies solar power can be converted or concentrated and turned into electricity needed. The process of converting or concentrating the solar power is essential in order to produce the electricity, because the density at which the solar radiation reaches the Earth is not sufficient for generating the thermodynamic cycle that is required for electricity production.[22]

Based on the way the electricity is being produced, whether it is by converting or by concentrating the solar power, the solar energy technologies are divided into different types. Photovoltaic solar cells are made to produce electricity by converting sunlight directly into the required electric charge. Electricity can also be produced using concentrating solar power, where electric charge appears from the turbine, which is driven by steam created from the heat, which is focused to a single point using solar towers or parabolic solar collectors. This technology allows producing the electricity during the absence of sunlight by storing the heat collected to maintain the power.

---

[20] **European Commission**, *"Renewables Make the Difference"*, 2011, p.4-5, viewed on 23.06.2013, available online at URL: http://www.energy.eu/publications/Renewables-make-the-difference-2011.pdf
[21] **International Renewable Energy Agency**, *Renewable Energy Technologies: Cost Analysis Series. Solar Photovoltaics,* 2012, viewed on 15.06.2013, available online at URL:
http://www.irena.org/DocumentDownloads/Publications/RE_Technologies_Cost_Analysis-SOLAR_PV.pdf
[22] **European Commission**, *Renewables Make the Difference,* 2011, p.15-16, viewed on 23.06.2013, available online at URL: http://www.energy.eu/publications/Renewables-make-the-difference-2011.pdf

Meanwhile, photovoltaic plants also have the capability to be connected not only to the electricity grid but to batteries as well, where the energy can be stored.[23]

1.6. The use of the energy

Solar energy, coming from the world's primary energy source - the sun, as a clean energy can be used for production of electricity or heat. Converting solar energy into energy used for heating and cooling purposes allows fulfilling the needs of heating in buildings and industrial processes, applying the results of production for domestic hot water, solar-assisted cooling, swimming pools, etc. Even the least advanced solar thermal systems are capable of providing sufficient energy for a (at times even substantial) part of domestic hot water needs. Despite the fact of such systems being noticeably more productive in sunny climates, the efficiency of new technologies allow contributing to hot water or space heating in any location within the EU. Solar energy can also be used by technologies that operate in some way similar to refrigerator technologies, which allow using the energy for cooling in air conditioning systems with heat absorption.[24] And the electricity created from solar energy can be adjusted and used for all our electricity needs.

The fact of actually existing tendency and/or intentions to use the energy produced from the solar or any other renewable energy, in general, is proven by such companies like "Apple" announcing this year their plans to run their "iCloud" data centres exclusively on renewable energy. Supermarket "Walmart" also announced its intentions by the year 2020 to be running all of the stores only on the renewables. As an addition to that, company "Google" has installed a 1.7MW solar plant already in 2007. Another confirmation of existence and even growth of the mentioned tendency to use the solar and other renewable energies is the fact of such companies as Volkswagen, Nike, Renault, HSBC, PepsiCo and Sumitomo following the example set by the companies mentioned above.[25]

1.7. Future solar energy demand

With a goal of confirming the relevance of the solar energy sector in the EU and while attempting to justify the promotion of development of the sector and the need of solar energy in general, it is important to establish the possibility and approximate matter of its future demand.

---

[23] **European Commission,** *Renewables Make the Difference,* 2011, p.14, viewed on 23.06.2013, available online at URL: http://www.energy.eu/publications/Renewables-make-the-difference-2011.pdf
[24] **European Commission,** *Renewables Make the Difference,* 2011, p.14, viewed on 23.06.2013, available online at URL: http://www.energy.eu/publications/Renewables-make-the-difference-2011.pdf
[25] **Ernst & Young Global Limited,** *Renewable energy country attractiveness index: May 2013*, Issue 37, 2013, viewed on 12.06.2013, p.11, available online at URL:
http://www.ey.com/Publication/vwLUAssets/Renewable_energy_country_attractiveness_indices_-_Issue_37/$FILE/RECAI-May-2013.pdf

Based on whether it will increase or decrease, or possibly not even arise at all, we can determine whether it is reasonable to invest in the solar energy sector or not.

Based on the data reflected in the figure 1, the solar energy in the EU (including photovoltaics, solar thermal electricity, solar thermal heating and hot water) has been growing in the analysed period of time and is expected to continue the growth in even larger portions during the next 17 years. According to the publications of the European Commission, solar energy is set to produce electricity in increasing amounts during the next several years and, based on projections, the output of the electricity produced could triple in the time from 2004 to 2020. This can also be observed in the figure 1.[26]

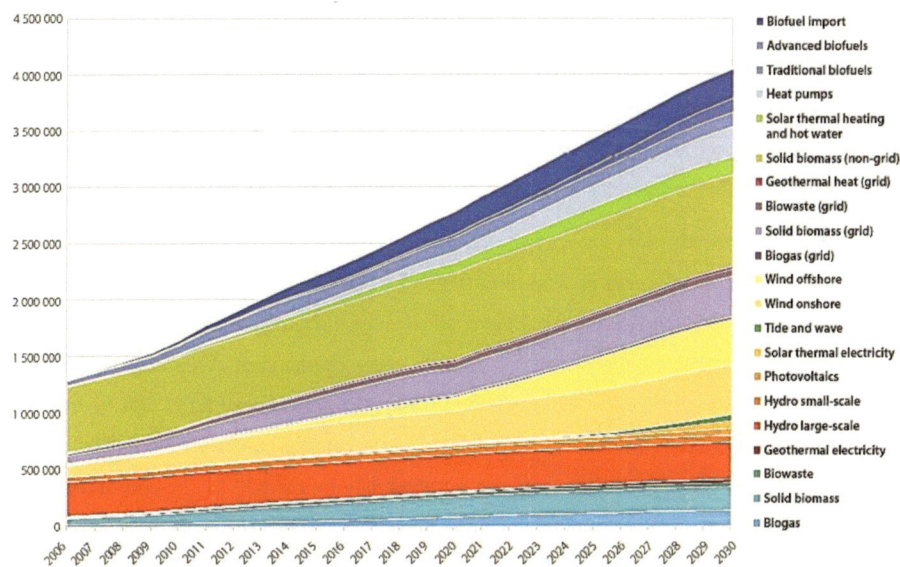

Fig 1. Estimate of renewable energy growth for the EU, 2006-2030, GWh

(Source: Green-X model from the Fraunhofer Institute and EEG (European Economics Group — Vienna University of Technology)

The demand of the electricity produced from solar energy can be affected by fluctuations of such economic indicators like gross domestic product (GDP) and gross value-added. GDP in the EU decreased by 0.9% in the last quarter of 2012 in comparison with the same period in 2011 and by 0.3% during the 2012 in general. It was the first year that had annual GDP decrease since 2009. Furthermore, according to the data of the last quarter of the 2012, gross value-added in construction and industry decreased significantly in comparison to the year before: construction by 4.7% and

---

[26] **European Commission,** *Renewables Make the Difference,* 2011, p.23, viewed on 23.06.2013, available online at URL: http://www.energy.eu/publications/Renewables-make-the-difference-2011.pdf

industry by 2.2%. This had to impact the consumption of electricity in the EU based on these two sectors being energy consumers of great importance.[27] Ergo, further fluctuations of these economic indicators can also have rather big influence to the demand of the solar energy.

Taking into consideration the fact of global economy facing problems, the market outlook becomes uncertain. Economic challenges in the government budgets as well as in the world economy can have a noticeable impact on the photovoltaics market. According to the data published by the International Renewable Energy Agency, European market of photovoltaics accounted for 80% of global demand in the last several years, which can have a substantial impact on supply and demand in the global industry of photovoltaics in case of any decrease in annual demand in the EU resulting from the economic depressions. Nevertheless, any decrease in the EU market could be compensated by boosts encouraged by policy measures in other photovoltaic markets such as China, Japan, India, Canada, the United States, Australia and other countries that show noticeable growth. The largest emerging markets recently are China, South Korea, the Middle East, India and other Southeast-Asian countries. The growth in these markets is considered to remain sustained, but it is not believed to be able to reach the levels of growth that was seen in the EU.[28]

During the recent years, the largest portion of the photovoltaics growth has been generated by the promotion policies, such as feed-in tariffs and other incentives, which allowed developing markets, reducing prices and raising the interest of investors. Given the rapid growth of installations during recent years, forecasting projections became really challenging. According to the International Renewable Energy Agency, the trends of 2011 set the projections of 131 GW to 196 GW of installed capacity in 2015. Moreover, the upper range is considered reasonable, while the lower range is considered to be too pessimistic and could already be reached by the end of 2013 or the beginning of 2014. Based on the photovoltaic roadmap created by the mentioned agency, by the year 2020 the installed power capacity of photovoltaics should result to 200 GW.[29]

1.8. The overview of the Solar Energy sectors of the EU countries

With the main aim of this study being the creation of an idea for a development model for the solar energy sector in Lithuania taking into account the experience of the EU, it is essential to screen out the countries of the EU with a most advances solar energy sector, while willing to have a

---

[27] **European Commission**, *Quarterly Report on European Electricity Markets*, March 2013, p.3, viewed on 16.06.2013, available online at URL:
http://ec.europa.eu/energy/observatory/electricity/doc/20130611_q1_quarterly_report_on_european_electricity_markets.pdf
[28] **International Renewable Energy Agency**, *Renewable Energy Technologies: Cost Analysis Series. Solar Photovoltaics*, 2012, viewed on 15.06.2013, available online at URL:
http://www.irena.org/DocumentDownloads/Publications/RE_Technologies_Cost_Analysis-SOLAR_PV.pdf
[29] **International Renewable Energy Agency**, *Renewable Energy Technologies: Cost Analysis Series. Solar Photovoltaics*, 2012, viewed on 15.06.2013, available online at URL:
http://www.irena.org/DocumentDownloads/Publications/RE_Technologies_Cost_Analysis-SOLAR_PV.pdf

good example of solar energy sector development. Since a number of the EU countries have noticeably more advanced solar energy sectors than Lithuania and, as a member of the EU, Lithuania is obliged to follow same rules and policies it makes the sectors comparable and allows applying similar measurements applied in the other EU member states.

1.9. Screening out the countries with the most advanced solar energy sector

Overall, in the EU, there is a potential seen of around 20 to 25 GW in the next few years, in case of using the right ways of support and promotion. In other case, the market has a high risk of collapsing to less than 10 GW per year. If such a turn of events occurs, it would affect the markets on a global scale with a substantial negative effect on the industry, companies bearing with low prices and general low demand. According to the data of the European Photovoltaic Industry Association, the most attractive solar energy sectors for investors are and will continue to be such countries of the EU like Germany, Italy, United Kingdom, Belgium and Greece. With the collapsed, due to the ups and downs of the economy, market in the Czech Republic and the Spanish market being also negatively affected, the two leading European markets remain to be Germany and Italy.[30]

Based on the data published in 2012 by KPMG International Cooperative, during the year 2011, Germany and Italy together accounted for nearly 50% of global photovoltaics installations. Both of these countries have used large-scale tariff reductions for their markets in order to attract the investors.[31]

One of the reasons of the European market shaping out into what is the current situation, with one countries leading in the amount of installed capacities of solar photovoltaics and others not being able to reach the similar level, is the speed of the development of technologies through the last several years. As reflected in the figure 2, Germany has the most developed market of solar photovoltaics, since it has experienced a steady growth for last several years. Some countries, such as Italy, Belgium and Czech Republic, experienced growth starting few years later than Germany, nonetheless, managed to reach high levels rather quickly. Spain appears to have not reached high results, since its market was affected negatively by the economic difficulties. The markets of the United Kingdom and France, by the results shown in this Figure, did not tap their potential.[32]

---

[30] **European Photovoltaic Industry Association**, *Global Market Outlook For Photovoltaics Until 2016*, 2012, viewed on 10.06.2013, available online at URL:
http://www.helapco.gr/ims/file/reports/Global%20Market%20Outlook%202016.pdf
[31] **KPMG International Cooperative**, *Green power 2012: The KPMG renewable energy M&A report*, 2012, p.11, viewed on 25.06.2013, available online at URL:
http://www.kpmg.com/CZ/cs/IssuesAndInsights/ArticlesPublications/Press-releases/Documents/KPMG-Green-Power-2012.pdf
[32] **European Photovoltaic Industry Association**, *Global Market Outlook For Photovoltaics Until 2016*, 2012, viewed on 10.06.2013, available online at URL:
http://www.helapco.gr/ims/file/reports/Global%20Market%20Outlook%202016.pdf

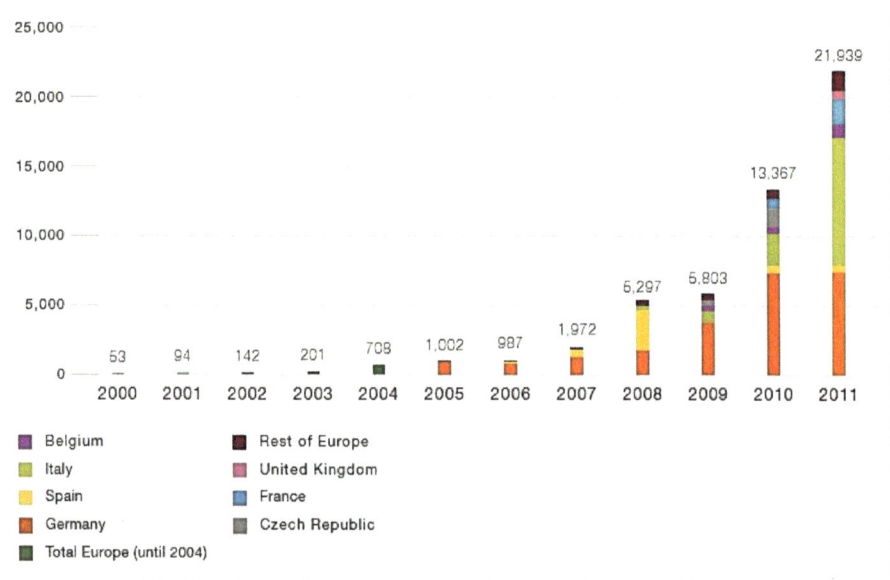

Fig 2. Evolution of European new grid-connected PV capacities
(Source: European Photovoltaic Industry Association, *Global Market Outlook For Photovoltaics Until 2016,* 2012, p.16)

For further analysis, examples and comparisons, based on the data collected, it is reasonable to choose Germany, Italy and United Kingdom, since these countries were mentioned most often as the ones with the most advanced and most experienced in the topic of solar energy. Germany and Italy, as mentioned earlier, are the two leading markets in the EU. For better understanding and comparisons, it is better to have more than 2 countries chosen for further analysis. Due to that, the United Kingdom was chosen, since it was considered as one of the most advanced markets and, although, in the figure 2, it is shown as the one that did not tap its potential, in the sources of 2013 it is described as one of the leaders, taking the 5th place in the world.[33]

1.10. The short overview of chosen countries and their solar energy sectors

As stated earlier, the chosen countries for the further analysis are Germany, Italy and the United Kingdom. The analysis is planned to include the general overview of the solar energy sectors of these countries as well as the main information about the policies and systems for renewables in general and other related aspects that help to describe the situation in the countries.

---

[33] **Ernst & Young Global Limited**, *Renewable energy country attractiveness index: May 2013*, Issue 37, 2013, p.16, viewed on 12.06.2013, available online at URL:
http://www.ey.com/Publication/vwLUAssets/Renewable_energy_country_attractiveness_indices_-_Issue_37/$FILE/RECAI-May-2013.pdf

Starting with the overview of the main aspects of the renewable energy policies and the legislative background for the solar energy sector and its development it has been chosen to begin by describing the situation in Germany. Then it is planned to characterize the situation in the United Kingdom and complete by defining the situation in Italy.

1.10.1. Germany

From the 1st of January of 2012 into effect came a new amendment of the Renewable Energy Sources Act (EEG) that was passed by the German Bundestag in June of 2011. This was already the third extensive revision, which was driven by the aims to make Germany's renewable energy more competitive. It introduces a market premium, which creates the conditions for electricity producers to sell the electricity generated from the renewable energies (including solar) on the electricity market of the country. This premium will function in parallel with the already existent fixed feed-in tariffs. The amendment also continues the system of a dynamic degression of tariffs for solar photovoltaics based on the increase of the capacity, which operates since 2010-2011.[34]

The major instrument in Germany used to support generating of electricity from the solar energy as well as other renewables is a feed-in tariff system. It applies for all technologies that are relevant in this field, except the ones that are used in combination with conventional power plants. This system provides fixed feed-in tariffs and from 2012 also as option - sliding feed-in premiums. The tariffs depend on the technology and capacity of the installation and are guaranteed for a period of 20 years. These tariffs are annually decreasing (by between 1.5 % and 24 % at the end of the year) for the new installations with the aim to stimulate technological learning. There is also the system of additional bonuses that are paid for distinct quality criteria, such as new technologies, high efficiency or auto-consumption, fulfilment of sustainability criteria. This model of support is not financed by the governmental budget, but by allocation to the final consumer, therefore, there is no general gap on it. It is required for operators to register their installations with the federal grid regulator in order to receive the financial support.[35]

The solar photovoltaic tariffs for 2012 in Germany are reflected in the table shown in the table 1. The most commonly in the EU supported installations of <30kW in Germany have support

---

[34] **Winkel Thomas, Rathmann Max, Ragwitz Mario, Steinhilber Simone, Winkler Jenny, Resch Gustav, Panzer Christian, Busch Sebastian, Konstantinaviciute Inga,** *Renewable Energy Policy Country Profiles,* 2012, viewed on 18.06.2013, available online at URL: http://www.ecofys.com/files/files/ecofys_re-shaping_country_profiles_2011.pdf
[35] **Winkel Thomas, Rathmann Max, Ragwitz Mario, Steinhilber Simone, Winkler Jenny, Resch Gustav, Panzer Christian, Busch Sebastian, Konstantinaviciute Inga,** *Renewable Energy Policy Country Profiles,* 2012, viewed on 18.06.2013, available online at URL: http://www.ecofys.com/files/files/ecofys_re-shaping_country_profiles_2011.pdf

level of 22.43 €cent/kWh. In general, the realizable photovoltaics potential in Germany is accounted to be 41.884.056 GWh by the year 2020 and 59.919.084 GWh by the year 2030.[36]

Table 1. PV tariffs for 2012 in Germany

Solar PV	Support level (€cent/kWh)
<30kW	24.43
30kW -100kW	23.23
>100kW	21.98
>1000kW	18.33
Ground-mounted installations	17.94-18.76

(Source: Winkel Thomas, Rathmann Max, Ragwitz Mario, Steinhilber Simone, Winkler Jenny, Resch Gustav, Panzer Christian, Busch Sebastian, Konstantinaviciute Inga, *Renewable Energy Policy Country Profiles,* 2012)

1.10.2. The United Kingdom

The major support instruments for the generation of electricity from renewable energies in the United Kingdom, at the national level, are the feed-in tariff system, which was introduced relatively recently, and the system of the renewables obligation. The installations can receive support based on their capacity: the ones less than 50 kW are only allowed apply for the feed-in tariff support system and the ones that have capacity between 50 kW and 5 MW can choose the support system of the renewables obligation or the feed-in tariff. The share of electricity produced from renewable energies increased from 1.8% to 6.8% in the time period from 2002 to 2010 using the support system of the renewables obligation. The system was, on the other hand, criticized for leading to the situation, where electricity generations get unreasonably high profits, electricity consumers experience the increase of costs and the deployment of the renewable energy electricity is increasing insufficiently. Due to that, the government of the United Kingdom recently suggested to revoke the system of renewables obligation from 2017 and, in order to cover all electricity generation from the renewables, expand the system of feed-in tariffs.[37]

The feed-in tariffs system in the United Kingdom was launched in 2010. It is targeted at the projects of the capacity up to 5 MW and granted for 20 to 25 years, based on the technology. Starting from September of the 2011, there have been nearly 100 000 installations, which were registered for the support, and, most importantly, 97% of those installations were solar energy

---

[36] **Winkel Thomas, Rathmann Max, Ragwitz Mario, Steinhilber Simone, Winkler Jenny, Resch Gustav, Panzer Christian, Busch Sebastian, Konstantinaviciute Inga**, *Renewable Energy Policy Country Profiles,* 2012, viewed on 18.06.2013, available online at URL: http://www.ecofys.com/files/files/ecofys_re-shaping_country_profiles_2011.pdf
[37] **Winkel Thomas, Rathmann Max, Ragwitz Mario, Steinhilber Simone, Winkler Jenny, Resch Gustav, Panzer Christian, Busch Sebastian, Konstantinaviciute Inga**, *Renewable Energy Policy Country Profiles,* 2012, viewed on 18.06.2013, available online at URL: http://www.ecofys.com/files/files/ecofys_re-shaping_country_profiles_2011.pdf

projects. Due to that, the government came to the decision to lower the tariffs for solar energy installations from 50 kW capacities and increase the tariffs for other renewables, since the government was concerned that the solar projects are taking a disproportioned share of funding. The table 2 reflects the tariffs for 2011 and 2012 that are effective since August of 2011 for solar energy installations. The tariffs will be annually reduced for new installations due to the reducing costs of the technologies. In general, the realizable potential of photovoltaics for the United Kingdom by the year 2020 is 12.976.848 GWh and 49.012.020 GWh for the year 2030.[38]

Table 2. PV tariffs for 2011 and 2012 in the United Kingdom

Solar Photovoltaics	Tariff (€cent/kWh)*
<4kW (new build)	43.5
<4kW (retrofit)	49.8
4-10kW	43.5
10–50kW	37.8/
50-100	37.8/21.9
100-150	35.3/21.9
150-250kW	35.3/17.3
>250kW	35.3/9.8
Stand-alone system	35.3/9.8

* Exchange rate used £1: €1.15

(Source: Winkel Thomas, Rathmann Max, Ragwitz Mario, Steinhilber Simone, Winkler Jenny, Resch Gustav, Panzer Christian, Busch Sebastian, Konstantinaviciute Inga, *Renewable Energy Policy Country Profiles*, 2012)

1.10.3. Italy

Directive 2009/28/EC, mentioned earlier in this study, in Italy was transposed into Italian law by the Legislative Decree 28/2011, which was approved by the Italian Government in March of 2011. The decree introduces some changes in the administrative procedures in the country concerning renewable energies and reforms the system of incentives for the renewables in general. This Italian system of incentives for the electricity generated from renewable energies is based on several support schemes such as: feed-in tariffs, tradable green certificates, scheme for photovoltaics (Conto Energia), scheme of simplified energy selling circumstances and scheme of net metering mechanism.[39]

---

[38] **Winkel Thomas, Rathmann Max, Ragwitz Mario, Steinhilber Simone, Winkler Jenny, Resch Gustav, Panzer Christian, Busch Sebastian, Konstantinaviciute Inga**, *Renewable Energy Policy Country Profiles*, 2012, viewed on 18.06.2013, available online at URL: http://www.ecofys.com/files/files/ecofys_re-shaping_country_profiles_2011.pdf
[39] **Winkel Thomas, Rathmann Max, Ragwitz Mario, Steinhilber Simone, Winkler Jenny, Resch Gustav, Panzer Christian, Busch Sebastian, Konstantinaviciute Inga**, *Renewable Energy Policy Country Profiles*, 2012, viewed on 18.06.2013, available online at URL: http://www.ecofys.com/files/files/ecofys_re-shaping_country_profiles_2011.pdf

The system of the feed-in tariffs is targeted at the renewables with maximum power output of 1MW and is considered as an alternative to the system of tradable green certificates (certificates are issued by the Italian Energy Service Provider). Tradable green certificates are targeted to support installations of more than 50 MWh and can be exchanged through bilateral contracts or traded on a market, which is managed by the Italian Electricity Market Administration. Both these systems grant the support for 15 years.[40]

Table 3. PV tariffs for 2012 in Italy

Solar Photovoltaics	1st semester (€cent/kWh)		2nd semester (€cent/kWh)	
	integrated	other	integrated	other
1 - 3 kW	27.4	24	25.2	22.1
3 - 20 kW	24.7	21.9	22.7	20.2
20 - 200 kW	23.3	20.6	21.4	18.9
200 - 1000 kW	22.4	17.2	20.2	15.5
1000 - 5000 kW	18.2	15.6	16.4	14
>5000 kW	17.1	14.8	15.4	13.3

(Source: Winkel Thomas, Rathmann Max, Ragwitz Mario, Steinhilber Simone, Winkler Jenny, Resch Gustav, Panzer Christian, Busch Sebastian, Konstantinaviciute Inga, *Renewable Energy Policy Country Profiles,* 2012)

Solar photovoltaics are being supported by the system of premiums, so called Conto Energia, which was initially launched in 2005 and then modified a number of times till the current stand. This system has been excessively successful due to the fact of 11.7 GW of capacity installed since its launch. The amount of the premium in based on size of installations and priority of ground installations against rooftop installations. There is a specific tariff system included for the installations with innovative technologies. This system grants the support for 20 years. In the table shown in table 3 we can see the detailed tariffs for solar photovoltaics in Italy for the year 2012. In general, the realizable potential of photovoltaics by the year 2020 is believed to be 11.127.996 GWh and 48.930.624 GWh by the year 2030.[41]

1.11. Choosing the country for further examples and comparisons

Based on the overview of the chosen countries and the additional information, which is shown in the figures 3 and 4 that reflect the most recent data of 2013, it is rational to choose Germany for further examples and comparisons.

---

[40] **Winkel Thomas, Rathmann Max, Ragwitz Mario, Steinhilber Simone, Winkler Jenny, Resch Gustav, Panzer Christian, Busch Sebastian, Konstantinaviciute Inga**, *Renewable Energy Policy Country Profiles,* 2012, viewed on 18.06.2013, available online at URL: http://www.ecofys.com/files/files/ecofys_re-shaping_country_profiles_2011.pdf
[41] **Winkel Thomas, Rathmann Max, Ragwitz Mario, Steinhilber Simone, Winkler Jenny, Resch Gustav, Panzer Christian, Busch Sebastian, Konstantinaviciute Inga**, *Renewable Energy Policy Country Profiles,* 2012, viewed on 18.06.2013, available online at URL: http://www.ecofys.com/files/files/ecofys_re-shaping_country_profiles_2011.pdf

According to the overview of the countries, Germany appears to have the most sophisticated system, which creates better conditions for the development of solar energy sector. Furthermore, the realizable potential for solar photovoltaics of Germany is larger than the potential of other analysed countries. Comparing the tariffs for photovoltaics in all the analysed countries, however, it is noticeable that the highest tariffs are in the United Kingdom while Germany and Italy have lower tariffs that have not a significant difference from one another. Nonetheless, Germany appears to lead in general, therefore it is chosen for further examples.

Rank	Country	RECAI score	Macro drivers			Energy market drivers			Technology-specific drivers		
			Macro stability	Ease of doing business	Total	Prioritization of renewables	Bankability of renewables	Total	Wind	Solar	Other technologies
1	US	71.6	77.3	71.2	73.6	38.6	75.0	60.4	68.0	79.4	52.0
2	China	70.7	66.4	44.2	53.1	58.3	62.9	61.1	77.5	78.5	55.7
3	Germany	67.6	75.7	61.0	66.9	55.4	72.7	65.8	59.9	63.0	45.8
4	Australia	60.6	85.1	73.0	77.9	53.6	65.6	60.8	46.9	57.2	30.1
5	UK	60.0	77.8	75.0	76.1	51.2	68.7	61.7	59.0	38.7	35.3
6	Japan	59.4	77.1	59.8	66.7	45.1	69.9	60.0	44.3	57.1	49.8
7	Canada	57.8	81.3	74.0	76.9	47.8	61.6	56.1	52.4	45.6	45.4
8	India	54.9	52.1	37.3	43.2	58.8	49.3	53.1	52.2	61.0	44.9
9	France	54.0	70.4	60.8	64.6	42.0	60.6	53.1	47.0	49.3	39.3
10	Belgium	53.9	67.3	78.0	73.7	65.0	61.1	62.6	42.4	36.9	26.4
11	Italy	53.7	46.7	45.7	46.1	50.7	63.9	58.6	37.9	57.9	42.8
12	South Korea	52.7	65.8	60.9	62.9	63.3	55.7	58.7	40.1	42.0	40.8

Fig 3. Renewable energy country attractiveness index scores and rankings at May 2013

(Source: Ernst & Young Global Limited, *Renewable energy country attractiveness index: May 2013*, Issue 37, 2013, p.16)

Concerning the additional information - the figure 3 - it reflects the data published by "Ernst & Young Global Limited" in May of 2013, where countries of the world are ranked according to their renewable energy markets and infrastructures, their suitability for individual technologies and projects. Moreover, figure 4 reflects the technology specific ranking of the countries, published by the same source basing the indices on a weighted average across energy market as well as macro and counting in the technological parameters. Based on this data, the country that has the most technologically advanced solar photovoltaics industry in the European Union is Germany, followed by earlier overviewed EU countries: Italy and UK as well as France and Belgium at some aspects, although, while specifying on the solar photovoltaic sector, the last two appear to be lower in the ranking system.

Rank	Onshore wind		Offshore wind		Solar PV		Solar CSP		Biomass	
1	US	71.0	UK	70.8	US	74.1	US	71.7	Germany	65.9
2	China	70.8	Germany	68.2	China	72.6	Australia	63.8	UK	62.1
3	Germany	66.6	China	65.7	Germany	72.5	Spain	60.7	China	61.6
4	UK	64.7	Belgium	59.5	Japan	66.6	China	60.4	US	61.4
5	Australia	63.2	Denmark	58.9	Australia	65.2	Chile	58.9	Brazil	61.0
6	Canada	63.2	US	57.5	Italy	61.5	India	58.7	Japan	60.0
7	Ireland	59.9	Netherlands	56.9	Canada	59.6	Israel	58.3	Belgium	59.0
8	Sweden	59.7	Sweden	55.9	India	58.7	Morocco	58.0	Sweden	58.4
9	India	59.7	Japan	53.0	UK	57.9	South Africa	55.4	Denmark	57.7
10	Denmark	59.4	Finland	52.1	France	57.6	Peru	52.0	Netherlands	57.2
11	Norway	59.2	South Korea	52.0	South Korea	57.1	Italy	52.0	Finland	57.2
12	Japan	58.7	Canada	49.5	Belgium	56.8	Saudi Arabia	50.4	South Korea	56.4
13	Netherlands	58.3	France	47.9	Israel	56.1	Brazil	49.8	Austria	55.0
14	Belgium	58.0	Norway	46.3	Spain	55.3	Turkey	49.0	France	54.3
15	France	57.9	Australia	44.9	Thailand	54.3	France	48.2	Canada	54.3
16	Poland	57.4	Ireland	43.8	Saudi Arabia	54.1	Portugal	48.0	Poland	53.7

Fig 4. Technology-specific indices and ranking of the countries
(Source: Ernst & Young Global Limited, *Renewable energy country attractiveness index: May 2013*, Issue 37, 2013, p.38)[42]

As a conclusion, the best country of the European Union for further examples and analysis, according to all the gathered information and results of the overviews, is Germany. Based on collected facts, the ranking results and the support systems and policies, level of technology development and realizable photovoltaics potential, the most experienced and advanced solar (photovoltaics) energy sector and market of the EU is developed in Germany. Due to this, it can be used as a good example for the currently developing solar energy sectors and markets in other EU countries, such as Lithuania.

---

[42] **Ernst & Young Global Limited**, *Renewable energy country attractiveness index: May 2013*, Issue 37, 2013, p.38, viewed on 12.06.2013, available online at URL:
http://www.ey.com/Publication/vwLUAssets/Renewable_energy_country_attractiveness_indices_-_Issue_37/$FILE/RECAI-May-2013.pdf

2. THE OVERVIEW OF THE CHOSEN COUNTRIES

Aiming to create a model or suggest a certain list of actions for solar energy sector development in Lithuania based on the example of the chosen country's solar energy sector development, in this case it would be Germany's example, it is very important to make an overview of these two countries, mostly concentrating on the sectors of renewable energy, solar energy and the business conditions in these countries. This will allow having a better understanding of the development of a relevant sector of the countries as well as screening out the relevant issues and possible mistakes that resulted to slowdowns in the sector development.

Lithuanian solar energy sector being the main focal point of this study and having shortly introduced the solar energy sector of Germany in previous chapter, in this part of the study it is important to make an overview of both countries with a bigger focus on Lithuania.

2.1. Case: Lithuania

Starting from the spring of 2004 Lithuania is a member state of the European Union. Therefore, Lithuania is obligated to meet the energy requirements set in the earlier mentioned Directive. Due to that during the recent decade Lithuanian renewable energy sector has had some positive changes while restructuring of the sector. This has been called a significant progress, since Lithuanian society as well as politicians were highly sceptical toward renewable energy for years and it stood in a way of any development of the sector. Taking this into consideration, it is understandable, why to this day in Lithuania renewable energy sources are mostly used in small systems and the sector is lacking of local qualified staff and authorities.[43]

According to the assessment of the administrative procedures in Lithuanian renewable energy sector published in 2012 by the European Commission, Lithuania has made some actions with the effort to simplify the procedures for small-scale solar (as well as biogas and wind) installations. In order to have a solar energy plant project approved by the municipality, by the simplified procedure in place since 2009, the plan has to be submitted to municipality and the approval or a reasoned rejection has to appear within 20 days from the submission date. In case of the absence of the respond in the set period of time there is a right to compensation of the resulting damages. Similar conditions of the permit to generate electricity were set for the smaller scale solar plants as well - the respond has to occur within the period of 30 days. Nonetheless, the results of the assess-

---

[43] **Sapronaitytė Inga**, *Opinion: Will renewable energy sources lead towards energy security?*, 12th June 2013, viewed on 08.07.2013, available online at URL:
http://www.lithuaniatribune.com/41337/opinion-will-renewable-energy-sources-lead-towards-energy-security-201341337/

ment indicate transparency and attitude issues since there are several different authorities involved in the procedures. This creates obstacles in the development of the sector.[44]

Based on the data provided by the Lithuanian confederation of renewable resources, there is a high potential for the renewable energy sector development in Lithuania. The Parliament of the Republic of Lithuania in 2011 passed a law, which sets the main measures and conditions for the promotion of renewable energies. As the key measure for promotion of the development of renewable energy the law determines the support for the purchase of renewable energy. Prices for the purchase of renewable energy are determined annually by the National Control Commission for Prices and Energy based on the type of the energy and the capacity. The feed-in tariffs are valid for the production with up to 30 kW of installed capacity. The fixed tariff is guaranteed by the law for 12 years. A good example of this promotion is the support for solar energy in particular. - the price for the purchase of the electricity produced from solar energy plant with a capacity up to 30 kW for the year 2012 was almost 10 times higher than the average market price: 1,80 Lt/kWh, when the market price in the country was around 18,46 ct/kWh (1 Lt = 100 ct = 0,28962213 Euros). The prices for the plants with bigger installed capacity will be determined by auctions (the producer with the lowest price offered wins) also for the period of 12 years. There is a limit set of the highest possible price for the auction for different plants, based on their capacity. Anyway, all the producers, no matter the capacity installed, are assured that all the electricity produced will be sold to the grid.[45]

Two other facts that serve as measures of promotion for the development of renewable energy sector are the law on excise, which considers electricity as a subject to an excise tax except for the electricity produced from renewables, and the reimbursement of costs that raise from the procedures of connecting the power plant to the grid. For the plants with the installed capacity of 30 kW or less the state is obliged to cover all 100 per cent of the connection expenses and for plants with installed capacity of more than 30 kW but not higher than 350 kW the state covers 20 per cent of expenses, while 40 per cent of expenses are to be covered for the plants with the installed capacity of more than 350 kW.[46]

---

[44] **Report for the European Commission**, *Renewable energy progress and biofuels sustainability*, September 2012, p.116, viewed on 07.07.2013, available online at URL:
http://ec.europa.eu/energy/renewables/reports/doc/2013 renewable energy progress.pdf
[45] **Poderis Justinas**, *Investment Opportunities in Renewable Energy Resources in Lithuania*, 5th of October 2012, viewed on 05.07.2013, available online at URL:
http://www.terralex.org/publication/p459bcb8ea3/investment-opportunities-in-renewable-energy-resources-in-lithuania
[46] **Poderis Justinas**, *Investment Opportunities in Renewable Energy Resources in Lithuania*, 5th of October 2012, viewed on 05.07.2013, available online at URL:
http://www.terralex.org/publication/p459bcb8ea3/investment-opportunities-in-renewable-energy-resources-in-lithuania

Further here, the figure 5 reflects the support to the renewable energy sector in years 2005 to 2012 - direct payments to producers of the electricity that was produced from renewable energies for the electricity supplied to the grid. It is obvious, that the law of 2011, which was mentioned earlier, basically doubled the amount support for the sector in 2011 compared to the numbers of 2010 and kept rising in following year. This shows a certain development in the sector - not only that the price for the electricity from renewable energies was enlarged, but actually the amount of electricity produced from renewable energy sources increased, while producers being encouraged by higher profits.

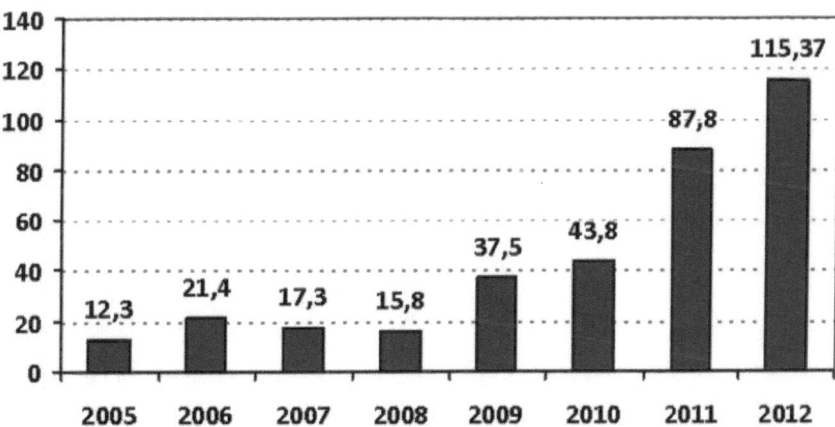

Fig 5. Direct payments to producers of electricity from renewable energy sources for electricity supplied to the grid in years 2005-2012 in millions of Litas (1 Litas = 0.289626196 Euros)
(Source: National Control Commission for Prices and Energy, *Commission approval of tariffs for electricity produced from renewable energy resources*, 2012)[47]

In table 4 the support for the solar energy sector is reflected in the electricity prices - the tariffs for electricity produced from photovoltaics that were set by the law for years 2011 and 2012. Prices are expressed in Euro, with an exchange rate of 1 Euro being worth of 3,45 Litas, for better comparison to the other support systems in the EU. The price for solar plants with up to 30 kW of the installed capacity is fixed, while prices for other capacities are meant as the highest possible prices that could be reached during the auction process. The years 2011 and 2012 mean the time when the solar energy plant is connected to the grid, since the price is being set depending on a year and capacity installed and then stays fixed for the 12 years, as it is set in the law. For more clearance of the system of support, the term "integrated" determines the solar energy plants that are built

---

[47] **National Control Commission for Prices and Energy**, *Commission approval of tariffs for electricity produced from renewable energy resources*, 2012, viewed on 11.07.2013, available online at URL: http://www.regula.lt/lt/naujienos/index.php?full=yes&id=14048

into the building and considered as a part of the building or a part of a roof construction. The tariff for such plants is higher mainly due to the difference in technology.

Table 4. PV tariffs for 2011 and 2012 in Lithuania

Solar energy plant	2011 (€/kWh)		2012 (€/kWh)	
	integrated	other	integrated	other
≤ 30 kW	0,47	0,47	0,52	0,42
30 - 100 kW	0,47	0,47	0,48	0,39
100 - 350 kW	0,45	0,45	0,37	0,30
> 350 kW	0,45	0,45	0,37	0,30
100 - 1000 kW	0,45	0,45	0,37	0,30
> 1000 kW	0,44	0,44	0,37	0,30

(Source: National Control Commission for Prices and Energy, *Commission approval of tariffs for electricity produced from renewable energy resources*, 2012)[48]

Alongside these efforts of promoting the development of the renewable energy sector there are still some legal barriers that stand in the way of investing in renewable energy. There have been no legislation in Lithuania introducing a long-term strategy for the development of the sector, which becomes a problem for investors to forecast the expansion or other changes of the sector as well as it makes it hard to anticipate the structural funds of European Union needed for the development of this field. Comparing to Germany, which in 2011 adopted a long-term policy anticipating a regular growth of the country's renewable energy sector and stating the result in 2050 to be the sector generating 80 per cent of country's energy, Lithuania presents no vision of its sector's future and, by this, creates economic and legal uncertainty in the development of the sector. As an addition to that serves a number of bureaucratic obstacles, such as the fact, that willing to invest in a solar plant, especially with bigger capacity than 30 kW, the investor has to get permits from not one, but several different institutions. This results to a delay of a project and creates uncertainties in planning and forecasting for investors. As a result, investors hesitate, whether it is a good idea to invest in solar energy or any other renewable energy in Lithuania for that matter.[49]

---

[48] **National Control Commission for Prices and Energy,** *Commission approval of tariffs for electricity produced from renewable energy resources*, 2012, viewed on 11.07.2013, available online at URL: http://www.regula.lt/lt/naujienos/index.php?full=yes&id=14048

[49] **Poderis Justinas,** *Investment Opportunities in Renewable Energy Resources in Lithuania*, 5th of October 2012, viewed on 05.07.2013, available online at URL: http://www.terralex.org/publication/p459bcb8ea3/investment-opportunities-in-renewable-energy-resources-in-lithuania

The assessment of renewable energy electricity policies and measures in Lithuania was presented in the report of 2012 for the European Commission. It stated that the policy commitments of the National Renewable Energy Action Plan (NREAP) were only partially fulfilled. Despite the fact that Lithuania adopted a number of new renewable energy electricity measures, it has failed to report on over 65 measures initially foreseen in the plan. The report does not reflect any clearance in what progress was achieved in Lithuania while preparing and adopting the measures and policies. On the other hand, the Renewable Energy Law adopted in 2011 reflects some of the planned measures, such as improvement of feed-in tariffs for renewable energy electricity, simplified procedures for issuance of permits for smaller installations, measures to guarantee power grid access and reduced rates for renewable energy power plants connection to the grid.[50]

The report also stated the adequacy of support levels for each renewable energy technology in Lithuania as fair. The feed-in tariffs scheme in Lithuania is technology-specific and the main support instruments are investment incentives financed by the structural funds of the European Union. Some of the tariffs paid were evaluated as not sufficient for some of the technologies. Furthermore, the long-term security of the support was also evaluated as fair. The feed-in tariffs are guaranteed for 12 years and the support scheme in general is set only till the end of 2020. As an addition to that, the scheme offers tendering, procedures of which in fact increase barriers for project developers, especially for smaller projects.[51]

2.1.1. Current situation

Since the adoption of the Renewable Energy Law in 2011 there have been several changes made in Lithuania regarding the support for solar energy power plants. According to the latest data available, the National Control Commission for Prices and Energy has published the new feed-in tariffs for photovoltaics for the 2nd and 3rd quarters of 2013. Comparing the new tariffs to the ones established in 2011 and 2012, the difference is obvious: the new tariffs are lower by approximately 50%. The new feed-in tariffs are shown in table 5. The prices originally are in Litas, but for the better comparison to other countries the tariffs are shown in Euro with an exchange rate of 1 Euro being worth of 3,45 Litas.

---

[50] **Report for the European Commission**, *Renewable energy progress and biofuels sustainability*, September 2012, p.116, viewed on 07.07.2013, available online at URL:
http://ec.europa.eu/energy/renewables/reports/doc/2013_renewable_energy_progress.pdf
[51] **Report for the European Commission**, *Renewable energy progress and biofuels sustainability*, September 2012, p.116, viewed on 07.07.2013, available online at URL:
http://ec.europa.eu/energy/renewables/reports/doc/2013_renewable_energy_progress.pdf

Table 5. PV tariffs for 3rd and 2nd quarter of 2013 in Lithuania

Solar energy plant	3rd quarter of 2013 (€/kWh)		2nd quarter of 2013 (€/kWh)	
	integrated	other	integrated	other
1 - 10 kW	0,23	0,18	0,28	0,22
10 - 100 kW	0,21	0,16	0,25	0,20
100 - 350 kW	0,19	0,15	0,23	0,19
> 350 kW	0,19	0,15	0,23	0,19

(Source: National Control Commission for Prices and Energy, *Commission approval of tariffs for electricity produced from renewable energy resources for 2013 3rd quarter*, 2013)[52]

After the former government of Lithuania established the high feed-in tariffs set in the Renewable Energy Law of 2011 it encouraged dozens of investors to invest in solar energy plants and, as a result, there were numbers of new power plants built all over the country in record time. But after the parliamentary elections in October of 2012 the leading position was taken by the Lithuanian Social Democratic Party, which this way took over the leadership from the centre-right coalition. The new government passed amendments to the Renewable Energy Law of 2011 slashing the feed-in tariffs significantly - for some technologies more than three-fold. The changes for solar energy plants with up to 30 kW of installed capacity were enacted by the new government in January of 2013 and later in May followed the changes for solar energy plants with bigger capacities.[53]

The actions of the new government towards the solar energy sector in Lithuania has angered a growing number of investors and, as claimed, has caused losses of around 40 million of US dollars in total. By latest available data, there have been 62 filed lawsuits against these actions of the government already and all demand compensations from the state. Due to the changes made by the new government it became nearly impossible for investors to forecast what the solar energy price will be after the solar energy plant will be launched. Therefore, investors are unable to plan and estimate their projects properly and the potential investors are highly likely to pull back.[54]

---

[52] **National Control Commission for Prices and Energy**, *Commission approval of tariffs for electricity produced from renewable energy resources for 2013 3rd quarter*, 2013, viewed on 11.07.2013, available online at URL: http://www.regula.lt/lt/naujienos/index.php?full=yes&id=49624
[53] **Jegelevicius Linas**, *Lithuanian solar investors sue government over FIT cuts*, 19th of June 2013, viewed on 19.07.2013, available online at URL:
http://www.pv-magazine.com/news/details/beitrag/lithuanian-solar-investors-sue-government-over-fit-cuts_100011774/#axzz2a0Tkwlda
[54] **Jegelevicius Linas**, *Lithuanian solar investors sue government over FIT cuts*, 19th of June 2013, viewed on 19.07.2013, available online at URL:
http://www.pv-magazine.com/news/details/beitrag/lithuanian-solar-investors-sue-government-over-fit-cuts_100011774/#axzz2a0Tkwlda

The most complains arise from the fact, that the feed-in tariffs were changed not only for the new solar energy plants, but also for the ones that were already submitted. Despite the fact that in the renewable energy law of 2011 the prices were granted for 12 years in advance the amended law reduces these prices this way the law being applied in reverse. The law affects over 15,000 persons, who submitted their requests to the Ministry of Energy for the needed permits in the years 2011 and 2012.[55]

2.1.2. Prospects and forecasts

According to the renewable energy country profile of Lithuania published in 2012 by a group of international researchers, the realizable potential of photovoltaics for Lithuania by the year 2020 is 4.41.864 GWh and 1.720.944 GWh by the year 2030.[56]

Despite the mentioned potential, after analysing the current situation in the solar energy sector in Lithuania and learning about the changes made by the new government it becomes hard to believe that Lithuanian solar energy sector will be able to reach the level in development, where the mentioned potential would be tapped. Furthermore, the renewable energy law of Lithuania states that the target for 2020 is to increase the capacity of solar energy to 10 MW, not counting the installations with the capacity up to 30 kW.[57] The target is lower than the potential and, due to the recent changes and disappointment of investors, the possibility of reaching higher capacities in later years is even smaller.

From conversations with several investors in solar energy sector of Lithuania, who invested in photovoltaic plants with the capacity of 30 kW (which were supposed to have fixed support for 12 years), it became clear that the investors are in fact disappointed by the changes in the support system and, moreover, confused and unsure about the future of their projects. Having started the project in the summer of 2012, the investors faced several obstacles, such as cancelation of all the permits issued by the Ministry of Energy in reverse (which affected the investors even though their permission was issued months before they got cancelled and all the investments where already made) and following changes. The permissions were returned to those, who later paid an additional deposit of 15,000 Litas to prove that they are capable of investing in their projects and building the solar plants. Moreover, there have been several changes in the tariffs for photovoltaics that changed

---

[55] **Jegelevicius Linas**, *Lithuanian solar investors sue government over FIT cuts*, 19th of June 2013, viewed on 19.07.2013, available online at URL:
http://www.pv-magazine.com/news/details/beitrag/lithuanian-solar-investors-sue-government-over-fit-cuts_100011774/#axzz2a0Tkwlda
[56] **Winkel Thomas, Rathmann Max, Ragwitz Mario, Steinhilber Simone, Winkler Jenny, Resch Gustav, Panzer Christian, Busch Sebastian, Konstantinaviciute Inga**, *Renewable Energy Policy Country Profiles*, 2012, viewed on 18.07.2013, available online at URL: http://www.ecofys.com/files/files/ecofys_re-shaping_country_profiles_2011.pdf
[57] **Renewable Energy Law,** 12th of May 2011, viewed on 23.07.2013, available online at URL: http://www3.lrs.lt/pls/inter3/dokpaieska.showdoc_l?p_id=398874

the price for the investors, which was initially written in the conditions for selling the electricity to the grid.[58]

All this only confuses investors and makes the sector much less attractive. Due to this it is highly unlikely for this sector to have big growth in the nearest future, unless the government creates a new strategy and makes favourable amendments to the law.

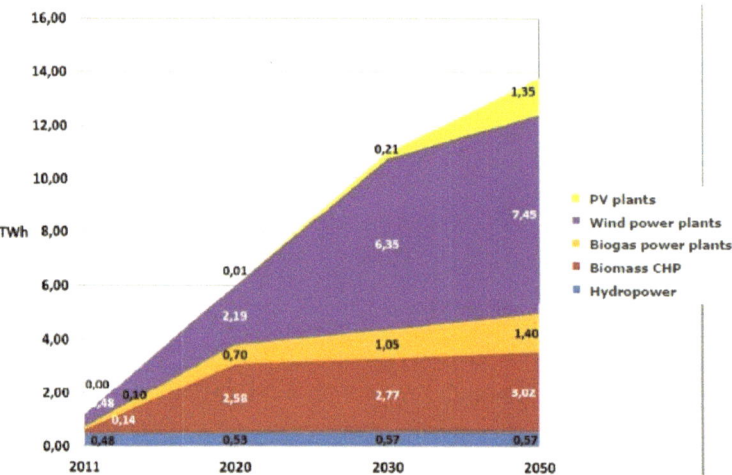

Fig 6. Forecast of power generation from renewable energies in Lithuania
(Source: Nagevicius Martynas, *Guidelines for new Lithuanian energy strategy*, 2013, p.16)

The figure 6 shows the forecast of growth in capacity of power produced from solar energy (PV plants) and other renewables in Lithuania in the period from 2011 to 2050. Besides hydropower, photovoltaics showing the least of the growth is understandable, considering the political framework in Lithuania for solar energy sector support. As mentioned earlier, due to unfavourable conditions that resulted from amendments made to the renewable energy law by the new government, dozens of investors experienced losses and the economical attractiveness to invest in solar energy in Lithuania decreased significantly. The greater growth can be only expected after making changes to the current situation in order to make the sector more attractive.

---

[58] **Maciulis V.**, *Saulės energetika Lietuvoje: Kas Gi Nutiko? (eng. Solar Energetics in Lithuania: What Happened?)*, 22nd of February 2013, viewed on 23.07.2013, available online at URL:
http://verslas.delfi.lt/energetika/vmaciulis-saules-energetika-lietuvoje-kas-gi-nutiko.d?id=60749505

2.1.3. Current conditions for Business

Estimating the current conditions for business in Lithuania creates a better understanding of the situation in the country as well as the environment for business and investing in the field of renewable energies based on the general background. In order to create a better understanding of the general conditions in the present time, the most comfortable it is to make an overview of the general and most relevant aspects published by the World Bank (figure 7).

REGION	Eastern Europe & Central Asia	DOING BUSINESS 2013 RANK	DOING BUSINESS 2012 RANK	CHANGE IN RANK
INCOME CATEGORY	Upper middle income	27	26	↓ -1
POPULATION	3,203,000	DOING BUSINESS 2013 DTF** (% POINTS)	DOING BUSINESS 2012 DTF** (% POINTS)	IMPROVEMENT IN DTF** (% POINTS)
GNI PER CAPITA (US$)	12,280	74.2	74.3	↓ -0.1

TOPIC RANKINGS	DB 2013 Rank	DB 2012 Rank	Change in Rank
Starting a Business	107	103	↓ -4
Dealing with Construction Permits	48	47	↓ -1
Getting Electricity	75	76	↑ 1
Registering Property	5	5	No change
Getting Credit	53	52	↓ -1
Protecting Investors	70	66	↓ -4
Paying Taxes	60	57	↓ -3
Trading Across Borders	24	26	↑ 2
Enforcing Contracts	14	15	↑ 1
Resolving Insolvency	40	40	No change

Fig 7. Ease of doing business in Lithuania 2013 (ranks out of 185 economies)
(Source: The World Bank, *Ease of Doing Business in Lithuania,* 2013)[59]

The figure 7 shows the fact sheet published by the World Bank in 2013, where Lithuania is ranked between 185 countries by the ease of doing business. In 2013 Lithuania is ranked 27th from 185 economies, which is a quite high value. By ease of starting business in Lithuania though, this economy was ranked as 107th. The ease of registering property seems to be ranked the highest (5th from 185) and investors' protection as well as electricity sector clearance has the lowest ranks.

Investment climate in Lithuania is claimed to include investor confidence in business security due to the European Union standards-compliant legislation that governs the investment environ-

---

[59] **The World Bank,** *Ease of Doing Business in Lithuania,* 2013, viewed on 26.07.2013, available online at URL: http://www.doingbusiness.org/data/exploreeconomies/lithuania

ment in general economic sector. Since 2011, the Government of Lithuania is making efforts to reduce the administrative barriers for businesses, optimize inspecting institutions, review business licensing, simplify business regulation, improving the system of consulting and competition rules. Local and foreign investors are presented by equal rights and conditions, as it is set in the law on investment.[60]

In general business conditions in Lithuania are considered to be above the average. On the other hand, analysing conditions for business in the solar energy the situation becomes less favourable at the time. With all changes made by government, there is an uncertainty about the future and investing looks more risky. Based on the situation in this sector, there is a possibility that the country in general might be considered as risky for investment in other sectors as well. Since Lithuania is not widely known for investment possibilities and security, what already makes investors hesitate, the sudden changes in the law and unpredictability of the future might make them hesitate even more or at all avoid investing in this country.

2.2. Case: Germany

In Germany photovoltaic systems are supported by the Federal Government in a way that allowed the solar energy sector and the renewable energy sector in general to develop better than in any other country of the European Union. Aiming to achieve reducing green gas emissions 80-95% by 2050 Germany is highly committed to developing the renewable energy sector. To this day Germany has a big success in progress that makes it to be the leading country in the EU in field of renewable energies.[61]

Starting the way to success, in 1991, German politicians passed the Renewable Energy Law (*"Erneuerbare Energien Gesetz" or EEG*) that became a measure for renewable energy support with long-lasting consequences. This law granted, mostly in Bavaria, a market for electricity produced from small hydroelectric power generators. The law made all utility companies obliged to plug all renewable energy plants, from the smallest to the biggest, into the national electricity grid as well as to buy the electricity produced by these plants at a fixed price, which was set to be higher than the market price in order to guarantee a return on investments of the producers over the long term.[62]

---

[60] **Energy Charter Secretariat,** *Investment Climate and Market Structure Review in the Energy Sector of Lithuania,* 2013, p.11, viewed on 19.07.2013, available online at URL:
http://www.encharter.org/fileadmin/user_upload/Publications/Lithuania_ICMS_2013_ENG.pdf
[61] **Epsea.org,** *Germany Invest on Alternative Energy: Leading to Make the World Greener,* 2011, viewed on 18.07.2013, available online at URL:
http://www.epsea.org/germany-invest-on-alternative-energy-leading-to-make-the-world-greener.html
[62] **Curry Andrew,** *Can You Have Too Much Solar Energy?,* March of 2013, viewed on 20.07.2013, available online at URL:
http://www.slate.com/articles/health_and_science/alternative_energy/2013/03/solar_power_in_germany_how_a_cloudy_country_became_the_world_leader_in_solar.html

In 2000, into effect came the German Renewable Energy Sources Act (also referred to as EEG), which influenced adoption of similar laws by many other countries all around the world. Since its adoption by Germany, there has been several amendments done that triggered a boom in solar energy sector. During a period of just few years, the annual revenue of the solar energy sector has increased to over 8.7 billion Euros. This has shown the creation of such political framework conditions being the main reason for success.[63]

The EEG principle is based on private and institutional investors in solar energy plants receive a granted feed-in tariff for the electricity produced and fed into the public grid. The prices for which the electricity is being bought are calculated in such way that investments made for installations of solar energy plants would become economically attractive. The fixed tariff depends on the capacity installed and the technology of the solar plant. The EEG guarantees fixed tariffs for period of 20 years, which creates secure conditions for investors to plan and invest in solar energy projects. Furthermore, electricity produced from solar energy sources, as from renewable energies in general, is given a status of priority, which means that it receives special conditions in order to be connected to the grid. This is because the law requires grid system operators to extend their grid in order to accommodate the connection of new solar energy capacities to the grid.[64]

Given these conditions, investors began to approach solar energy as long-term investments. The bureaucratic barriers were minimized and paperwork for all the renewables was streamlined. German companies became the world's leading companies in solar energy technologies and research and are still holding that status nowadays. In May 2013, on one sunny day, Germany generated 22 GW of energy from the solar power, which accounts for half of the world's total and equals the energy produced by 20 nuclear power plants. Due to competition in the market (Germany and elsewhere) the price for solar panels has decreased by 66 per cent since 2006, what, according to some studies, in few years will let the cost of energy generated from solar power compete with energy production from coal.[65]

---

[63] **German Solar Industry Association,** *Policy framework: The Renewable Energy Sources Act (EEG),* 2013, viewed on 18.07.2013, available online at URL: http://www.solarwirtschaft.de/en/photovoltaic-market/political-framework.html
[64] **German Solar Industry Association,** *Policy framework: The Renewable Energy Sources Act (EEG),* 2013, viewed on 18.07.2013, available online at URL: http://www.solarwirtschaft.de/en/photovoltaic-market/political-framework.html
[65] **Curry Andrew,** *Can You Have Too Much Solar Energy?,* March of 2013, viewed on 20.07.2013, available online at URL: http://www.slate.com/articles/health_and_science/alternative_energy/2013/03/solar_power_in_germany_how_a_cloudy_country_became_the_world_leader_in_solar.html

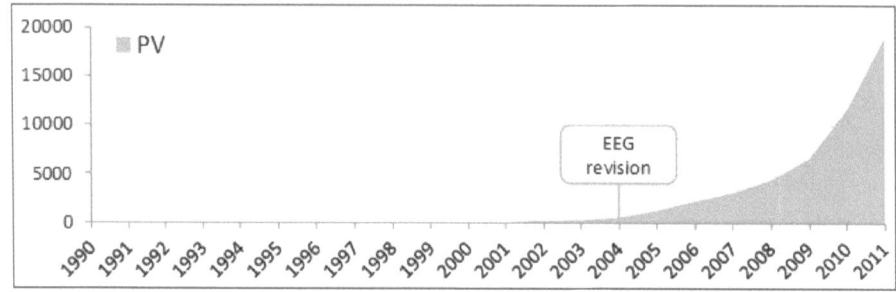

Fig 8. The impact of EEG revision on PV electricity production (in GWh)
(Source: Marquardt Jens, *Renewable Energy Projects - The German Energy Transition in a Multi-Level Perspective*, 2012)[66]

In figure 8 we can see an effect that one of the mentioned amendments, in this case - revision made in 2004, of the German Renewable Energy Sources Act had on the solar energy sector. The growth in installed capacities triggered by the law amendment is significant. This only proves the success of Germany in solar energy sector development.

2.2.1. Current situation

In the German progress report, where the assessment of the administrative procedures has been presented to the European Commission, it states that there have been no general non-cost barriers identified in renewable energy sector. All the existing procedures are being regularly checked and, if needed, improved. All the measures and conditions remain as planned and function successfully. Some variety occurs in different parts of Germany, such as only single permit needed or, for example, in Brandenburg, the online application is possible, etc.[67]

The assessment of policies and measures concerning electricity produced from renewable energy sources has shown that Germany has fully fulfilled all of the National Renewable Energy Action Plan policy commitments. Having fulfilled all the commitments, Germany has also implemented several new measures that were not foreseen in the NREAP. In 2011, Germany shut down 8 of its oldest nuclear power plants permanently. The remaining 9 reactors are planned to be shut down gradually till the year 2022. The needed capacity is planned to be replaced mostly by the

---

[66] **Marquardt Jens,** *Renewable Energy Projects - The German Energy Transition in a Multi-Level Perspective*, 2012, viewed on 21.07.2013, available online at URL:
http://www.academia.edu/2276998/Renewable_Energy_Projects_-_The_German_Energy_Transition_in_a_Multi-Level_Perspective
[67] **Report for the European Commission,** *Renewable energy progress and biofuels sustainability,* September 2012, p.113, viewed on 07.07.2013, available online at URL:
http://ec.europa.eu/energy/renewables/reports/doc/2013_renewable_energy_progress.pdf

renewable ("green") energies. In Germany there is a special term for this strategy - "Energiewende".[68]

In the report for the European Commission the adequacy of support for each renewable energy technology in Germany has been defined as good. Using technology-specific feed-in tariff system and optional premium for the support since 2012, Germany was claimed to have sufficient levels of support for all technologies (some even considered as quite high). Furthermore, the long-term security of support in Germany has also been defined as good, since the tariffs are guaranteed for 20 years and constant revisions and tariff adoptions are being used in order to avoid endangering the sector by increasing costs.[69]

The figure 9 shows the progress of solar energy sector in Germany since 2000 until 2013. Since the graph was created in July of 2013, the numbers for 2013 are not reflecting the result of the whole year.

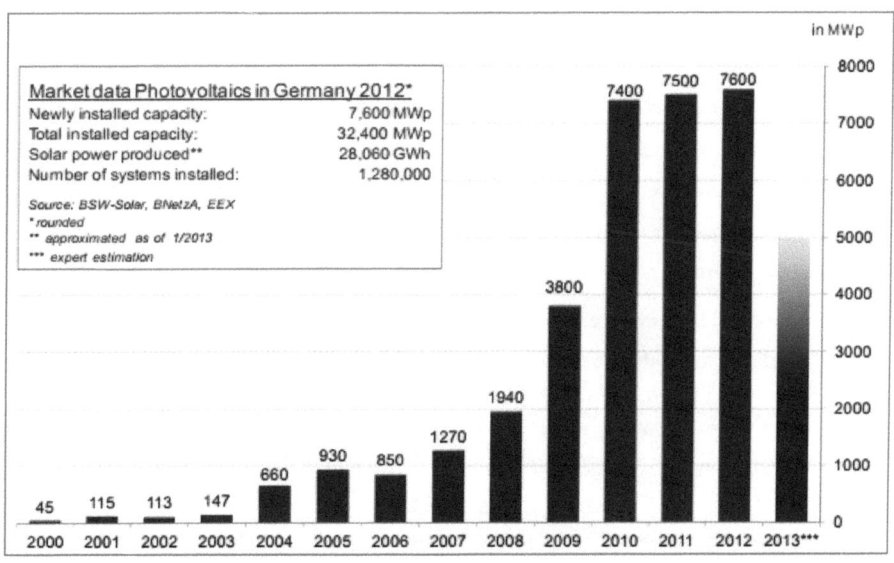

Fig 9. PV on the path to becoming a key pillar of a sustainable energy supply in Germany
(Source: German Solar Industry Association, Statistic data on the German solar power (photovoltaic) industry, July 2013, p.2)[70]

---

[68] Report for the European Commission, *Renewable energy progress and biofuels sustainability*, September 2012, p.113, viewed on 07.07.2013, available online at URL: http://ec.europa.eu/energy/renewables/reports/doc/2013_renewable_energy_progress.pdf
[69] Report for the European Commission, *Renewable energy progress and biofuels sustainability*, September 2012, p.113, viewed on 07.07.2013, available online at URL: http://ec.europa.eu/energy/renewables/reports/doc/2013_renewable_energy_progress.pdf
[70] German Solar Industry Association, *Statistic data on the German solar power (photovoltaic) industry*, July 2013, p.2, viewed on 20.07.2013, available online at URL:
http://www.solarwirtschaft.de/fileadmin/media/pdf/2013_2_BSW-Solar_fact_sheet_solar_power.pdf

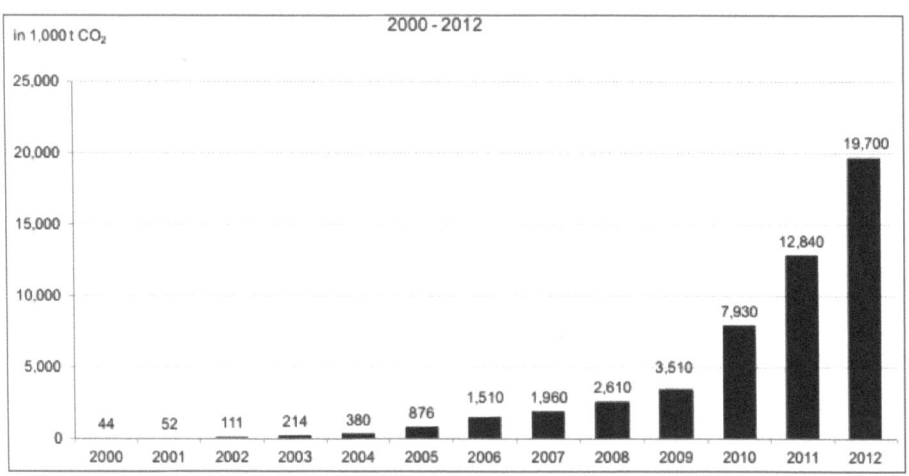

Fig 10. CO2 Savings through PV systems

(Source: German Solar Industry Association, Statistic data on the German solar power (photovoltaic) industry, July 2013, p.3)[71]

The figure 10 shows the savings of carbon dioxide through the usage of the solar energy systems in Germany. The data reflects the situation from 2000 till 2012. Based on earlier described German strategy, the numbers in this Figure are highly important for Germany's solar energy sector - it shows the progress on the way to fulfilling Germany's goal of reducing the greenhouse gas emissions to the minimum.

In the figure 11 the most recent tariffs for photovoltaics are reflected in a value of Euro cents per kWh. Tariffs are lower than in the previous year by approximately 4 cents.

Compensation in cents/kWh	Nominal array output			
Month of installation	up to 10 kW	up to 40 kW	Up to 1 MW	Up to 10 MW
May 2013	15.63	14.83	13.23	10.82
June 2013	15.35	14.56	12.99	10.63
July 2013	15.07	14.30	12.75	10.44

Fig 11. Most recent PV tariffs in Germany

(Source: Schlütersche Verlagsgesellschaft mbH & Co. KG, *German PV drops to 15 cents max*, 2nd of May 2013)[72]

---

[71] **German Solar Industry Association,** *Statistic data on the German solar power (photovoltaic) industry,* July 2013, p.3, viewed on 20.07.2013, available online at URL:
http://www.solarwirtschaft.de/fileadmin/media/pdf/2013 2 BSW-Solar fact sheet solar power.pdf
[72] **Schlütersche Verlagsgesellschaft mbH & Co. KG,** *German PV drops to 15 cents max*, 2nd of May 2013, viewed on 22.07.2013, available online at URL:
http://www.renewablesinternational.net/german-pv-drops-to-15-cents-max/150/510/62457/

2.2.2. Prospects and forecasts

Regarding the realizable photovoltaics potential in Germany - it has been accounted to be 41.884.056 GWh by the year 2020 and 59.919.084 GWh by the year 2030.[73] According to the overview of the country's solar energy sector and its policies, the success is already visible and stated as the biggest in European Union. Due to that it is reasonable to believe that Germany is highly capable to tap its potential or even set the higher potential after recounting in few years.

The German Renewable Energy Sources Act aims to enlarge the portion of electricity produced from renewable energy sources in overall electricity supply to approximately: 35 % until 2020; 50 % until 2030; 65 % until 2040; 80 % until 2050.[74]

Photovoltaic storage systems in Germany are forecasted for next five years to have a growth by nearly 100% per year. In 2017 it is expected to reach the level of approximately 7 GWh. Germany is forecasted to account for up to 70% of installed capacity of photovoltaic storage systems worldwide. It is expected to lead in this area for at least next 5 years.[75]

According to the International Energy Agency, German electricity prices to this day are among the highest in the European Union. Furthermore, this situation remains, while the wholesale prices are relatively low. According to the agency, this should be a warning signal for Germany to be prepared to manage or avoid the possible problems in the future. At this point, it is highly recommended to maintain the current policy framework and avoid initiating any changes in order to remain as strong as it is at the present. Sudden changes are highly likely to violate investor confidence and it will increase the costs in the long-term perspective.[76]

Key recommendations for the future solar sector of Germany are to create measures to ensure the costs of Energiewende to be as low as possible and allocated across customer groups, as well as limit the growth of renewable energy sector, while controlling it in order to avoid unexpected booms, slightly changing the prices to use all benefits from the rapid decrease of technology costs.[77]

---

[73] Winkel T., Rathmann M., Ragwitz M., Steinhilber S., Winkler J., Resch G., Panzer C., Busch S., Konstantinaviciute I., *Renewable Energy Policy Country Profiles* 2012, viewed on 18.06.2013, available online at URL: http://www.ecofys.com/files/files/ecofys re-shaping country profiles 2011.pdf

[74] **Federal Ministry for the Environment**, *Nature Conservation and Nuclear Safety, Renewable Energy Sources Act (EEG) 2012*, May. 2013, viewed on 26.07.2013, available online at URL:
http://www.erneuerbare-energien.de/en/unser-service/mediathek/downloads/detailview/artikel/renewable-energy-sources-act-eeg-2012/

[75] **Germany Trade & Invest**, *Industry Overview: The Photovoltaic Market in Germany*, Issue 2013/2014, p.8, viewed on 24.07.2013, available online at URL:
http://www.gtai.de/GTAI/Content/EN/Invest/_SharedDocs/Downloads/GTAI/Industry-overviews/the-photovoltaic-market-in-germany.pdf

[76] **International Energy Agency**, *IEA says further action is needed if Germany's Energiewende is to maintain a balance between sustainability, affordability and competitiveness*, 24 May 2013, viewed on 29.07.2013, available online at URL: http://www.iea.org/newsroomandevents/pressreleases/2013/may/name,38340,en.html

[77] **International Energy Agency**, *IEA says further action is needed if Germany's Energiewende is to maintain a balance between sustainability, affordability and competitiveness*, 24 May 2013, viewed on 29.07.2013, available online at URL: http://www.iea.org/newsroomandevents/pressreleases/2013/may/name,38340,en.html

2.2.3. Current conditions for Business

As well as for Lithuania, the conditions for business in Germany are chosen to be overviewed using the factsheet published by the World Bank in 2013. This allows following comparisons of the two countries to be more accurate.

REGION	OECD high income	DOING BUSINESS 2013 RANK	DOING BUSINESS 2012 RANK	CHANGE IN RANK
INCOME CATEGORY	High income	20	18	↓ -2
POPULATION	81,726,000			
GNI PER CAPITA (US$)	43,980	DOING BUSINESS 2013 DTF** (% POINTS) 77.3	DOING BUSINESS 2012 DTF** (% POINTS) 77.7	IMPROVEMENT IN DTF** (% POINTS) ↓ -0.4

TOPIC RANKINGS	DB 2013 Rank	DB 2012 Rank	Change in Rank
Starting a Business	106	100	↓ -6
Dealing with Construction Permits	14	12	↓ -2
Getting Electricity	2	2	No change
Registering Property	81	76	↓ -5
Getting Credit	23	23	No change
Protecting Investors	100	98	↓ -2
Paying Taxes	72	88	↑ 16
Trading Across Borders	13	11	↓ -2
Enforcing Contracts	5	6	↑ 1
Resolving Insolvency	19	12	↓ -7

Fig 12. Ease of doing business in Germany 2013 (ranks out of 185 economies)
(Source: The World Bank, *Ease of Doing Business in Germany*, 2013)[78]

The figure 12 reflects the factsheet with general information and the ranks for Germany on the ease of doing business in this country. In 2013, Germany was ranked 20th from 185 economies by the ease of doing business. Nonetheless, it was ranked 106th by the ease of starting business in the country. Germany is ranked 2nd from 185 economies by the ease of getting electricity, which is almost the highest rank overall and the highest rank for Germany. This shows German electricity sector to be well developed and have better conditions than most of other countries. The lowest, besides the ease of starting business, is the rank for protecting investors - 100th from 185 economies.

---

[78] **The World Bank**, *Ease of Doing Business in Germany*, 2013, viewed on 27.07.2013, available online at URL: http://www.doingbusiness.org/data/exploreeconomies/germany

2.3. Comparison of the analysed countries

In order to be able to compare the two countries, for the more accurate result, it was crucial to estimate both countries using mostly the same methods, same sources and giving the same conditions. With the aim of suggesting the solar energy sector development model for Lithuania, it is important to screen out the disadvantages of the current situation, especially comparing to the situation in Germany.

The table 6 allows us to compare the assessments of renewable energy electricity policies and measures that were reported after evaluating both countries. Here it is clearly visible that Germany results are at all points better than results of Lithuania. The differences in economy or weather or size of the country have no influence on why the results of Lithuania are worse. This assessment just shows that Lithuania did not handle the responsibility of fulfilling its commitments to the plan that was set by their own government, while Germany managed to fulfil their commitments.

On the other hand, the second evaluation reflected in this table for adequacy of support levels for each technology could have been partly influenced by the economic situation of countries. In this case, it is possible, that Germany is more capable of getting good results due to the stronger economy and better financial situation than in Lithuania.

Table 6. Assessment of RES-E policies and measures: Lithuania vs. Germany

	Lithuania	Germany
Fulfilment of National Renewable Energy Action Plan policy commitments	Partially	Yes
Adequacy of support levels for each technology	Fair	Good
Long-term security of support	Fair	Good

(Source: Report for the European Commission, *Renewable energy progress and biofuels sustainability*, September 2012)[79]

The most important part of this table is the evaluation for long-term security of support. Analysing the recent situation in the solar energy sector of Lithuania, there have been multiple sources found, where investors claim one of the most important reasons of the uncertainty of the

---

[79] **Report for the European Commission**, *Renewable energy progress and biofuels sustainability*, September 2012, viewed on 07.07.2013, available online at URL:
http://ec.europa.eu/energy/renewables/reports/doc/2013_renewable_energy_progress.pdf

sector to be the absence of the long-term strategy. To this day Lithuania has the plan till 2020, while Germany has a strategy, the long term policy adopted in 2010, which anticipates the renewable energy growth and estimates the measures and goals to be fulfilled by the year 2050. This is the guarantee for investors, that the sector has the future. As an addition to that, Lithuania stated by law to support solar energy with feed-in tariffs granted for 12 years, while Germany granted the support for 20 years, this way increasing the attractiveness and the security of investment.

Figure 13 reflects the ranking of the countries by the ease of doing business. The countries in the figure were chosen by the World Bank based on their comparability to one another. The figure shows Germany being ranked as 20th, while Lithuania is 27th. This shows that ease of doing business in Lithuania is worse than in Germany, but not with significant difference. Both countries are ranked higher than the average of the Eastern Europe and Central Asia region.

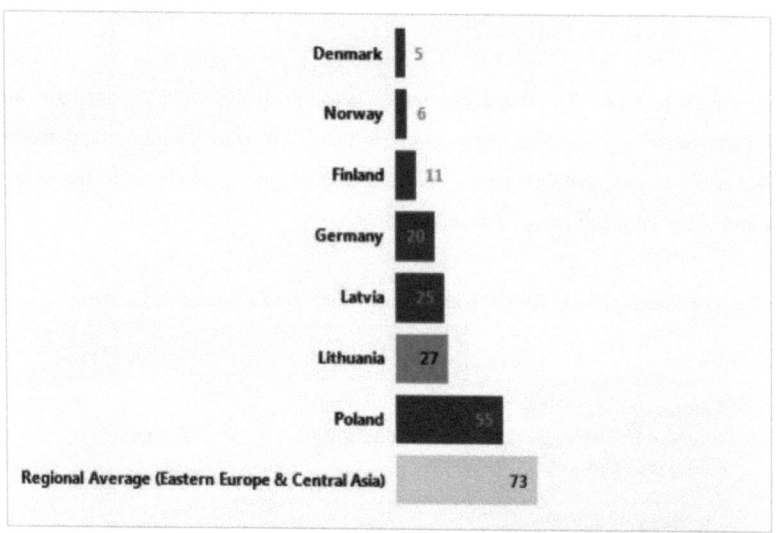

Fig 13. Lithuania and comparator economies ranked by ease of doing business
(Source: The World Bank, *Economy Profile: Lithuania, 10th Edition*, 2013, p.7)[80]

The ranking is based on the set of indicators measuring and benchmarking regulations that apply to small and medium-size domestic businesses from the beginning till the end of their existence.[81]

The figure 14 reflects the rankings, published by the World Bank, of the comparator economies on the ease of starting a business. In this ranking Lithuania and Germany are evaluated

---

[80] **The World Bank**, *Economy Profile: Lithuania, 10th Edition*, 2013, p.7, viewed on 27.07.2013, available online at URL: http://www.doingbusiness.org/~/media/giawb/doing%20business/documents/profiles/country/LTU.pdf
[81] **The World Bank**, *Economy Profile: Lithuania, 10th Edition*, 2013, p.7, viewed on 27.07.2013, available online at URL: http://www.doingbusiness.org/~/media/giawb/doing%20business/documents/profiles/country/LTU.pdf

approximately as equal: Lithuania ranked 107th, while Germany ranked 106th. This shows that in both countries it will take a similar set of conditions to create a company and start the activity. Both countries are ranked lower than the average of the region, which shows that conditions are less favourable than in most countries.

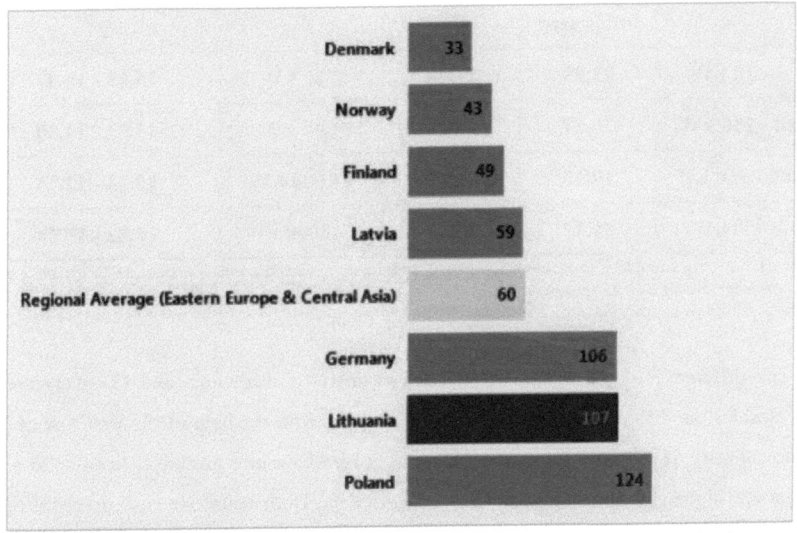

Fig 14. Lithuania and comparator economies ranked by ease of starting a business
(Source: The World Bank, *Economy Profile: Lithuania, 10th Edition*, 2013, p.16)[82]

The ranking is based on the average of rankings on main component indicators: paid-in minimum capital requirement, which is needed before registration of the company; procedures that are needed or commonly done by an entrepreneur to start up and operate business; time and cost needed for completing the required procedures.

While comparing the conditions in Lithuania and Germany also one of the most important aspects would be the support level for the solar energy in particular, evaluation of which was partially reflected in assessments shown in table 6. While numbers being the most understandable way of visualising and comparing the level of the support, the table 7 shows the feed-in tariffs of both countries for solar energy plants for most recent period of time. In order to achieve more accurate comparison the prices are expressed in same currency and detail. The exchange rate of 1 Euro is 3.45 Litas. All numbers are not rounded.

---

[82] **The World Bank**, *Economy Profile: Lithuania, 10th Edition*, 2013, p.7, viewed on 27.07.2013, available online at URL: http://www.doingbusiness.org/~/media/giawb/doing%20business/documents/profiles/country/LTU.pdf

Table 7. PV tariffs of Lithuania and Germany

Solar energy plant	Lithuania		Solar energy plant	Germany
	3rd quarter of 2013 (€cent/kWh)			May - July of 2013 (€cent/kWh)
	integrated	other		
1 - 10 kW	22.89	17.68	≤ 10 kW	15.63 - 15.07
10 - 100 kW	20.57	16.23	≤ 40 kW	14.83 - 14.30
100 - 350 kW	19.13	15.07	≤ 1000 kW	13.23 - 12.75
> 350 kW	19.13	15.07	≤ 10000 kW	10.82 - 10.44

(Sources: 1) National Control Commission for Prices and Energy, *Commission approval of tariffs for electricity produced from renewable energy resources for 2013 3rd quarter*, 2013; 2) Schlütersche Verlagsgesellschaft mbH & Co. KG, *German PV drops to 15 cents max*, 2nd of May 2013)

The difference of the prices in the feed-in tariffs of Lithuania and Germany are clearly visible. Besides the difference of the reflected capacities, with the help of the first line of prices at the same capacity, it is easily noticeable that prices for electricity produced from solar energy in Lithuania are higher by approximately 2 - 7 Euro cents. Both countries have lowered the prices several times from the day when the feed-in tariffs were first estimated, although Germany adopted the feed-in tariff system earlier than Lithuania and lowered the prices deliberately with difference of just around 5 Euro cents, when Lithuania lowered the prices in a very sudden way causing resentment among investors.

As a conclusion it is safe to say that in most of the aspects in general both countries seem to not have significant differences. On the other hand, Germany has better evaluations than Lithuania in very important aspects, although Lithuania has higher feed-in tariffs established for the recent present. Based on analysed sources of information, the long-term strategy is highly important to investors when assessing the attractiveness of investment into the solar energy. Germany is repeatedly mentioned as an example of a successfully functioning solar energy sector and the clear long-term strategy with favourable conditions and security. While the feed-in tariff in Lithuania is higher than in Germany, the security and certainty of an investment is claimed to be poor. Due to the unpredictability of government actions the sector became risky for investors. In order to further promote the development of the solar energy sector to increase the attractiveness of investing in solar energy there have to be certain measures brought into an action.

3. DEVELOPING THE MODEL FOR THE PROMOTION OF THE DEVELOPMENT OF THE SOLAR ENERGY SECTOR OF LITHUANIA

After analysing the situation in the solar energy sector of Lithuania, while comparing to the examples from the most advanced in the relevant field country in order to screen out the aspects that could be improved and create an understanding of what possible actions might be suitable and helpful in this case, the next rational step would be suggesting a model for solar energy sector of Lithuania that would include ideas for increasing the attractiveness of the sector.

The analysis of the examples from other countries that are considered as advanced in the field of renewable energy and a focused overview of measures and policies adopted in Germany helped to create a general understanding of the way the sector functions under different conditions and what measures create a positive effect on the sector attractiveness. The main goal was to understand what differed in Lithuania that was most important in the way of developing the sector and whether it could be improved in order to create positive changes in the development of the solar energy sector.

With all the differences that occurred during the process of analysing the solar energy sector of the chosen countries it was substantial to screen out the most essential differences and to create an idea with possible improvement model for the sector, where these differences would serve as a direction for the changes. The model would consist of steps that in this case are considered as recommendable actions needed in order to achieve the wanted result of increasing the attractiveness of the sector for investors.

3.1. The description of the model

To begin with, the model for the development of solar energy sector of Lithuania is aimed to increase the attractiveness of the sector for investors, while the development of the sector highly depends on the willingness of investors to invest in the solar energy technologies in the country. Investing in the solar energy increased the installed capacity in the territory of Lithuania, which in the future will result to the certain amount of electricity produced from solar energy in the country and ensure a certain level of security of the electricity supply. Furthermore, based on several sources, the cost of solar energy technologies decrease every year and the electricity production costs are one of the lowest in the field of energy production. Due to this the price of the electricity produced from solar energy will be able to compete with electricity prices of other sources in several years or even become one of the cheapest in longer perspective.

Willing to increase the security of electricity supply in the future, Lithuania should develop a strategy that foresees the development of renewable energy sector and solar energy sector in particular, since recently it suffered a decrease in the trust of investors. The strategy has to be set for a long term instead of the current plan that only covers the actions and measures for the period till 2020 and grants the feed-in tariff for 12 years.

3.1.1. Adopting the Long-Term Strategy

According to the results that occurred in the process of analysing the differences between the solar energy sector in Lithuania and the solar energy sector in Germany, the long-term strategy was adopted in Germany for a period till the year 2050 and mentioned in several cases as an example for Lithuania, since many investors claimed the absence of such strategy in Lithuania to be the indicator of a certain risk for investment and the cause of an uncertainty in the future of a development of the sector.

Based on the analysis, Lithuania has only established a plan for the period until 2020, which is considered as not sufficient for investors to be certain in the security of the sector in order to take a risk of investing. The main reason why in such situation dozens of investors decided to invest in the solar energy sector despite the absence of strategy and secure long-term perspectives - was the high feed-in tariff.

Due to the recent changes in the sector the tariffs were lowered and this way the main advantage for the investment disappeared. According to the analysis, many investors and other solar energy market participants no longer see the future in the sector and, without these advantages, no longer intend to invest. This can significantly slow down the development of the sector. In order to increase the attractiveness of the solar energy sector of Lithuania and promote its development the first step would be adopting a long-term strategy, as it was done by most of the currently leading industries.

The long-term strategy for solar energy sector development in Lithuania is needed for at least the period until 2050 as it shown in the example of the leading countries, such as Germany. The strategy has to reflect the solar energy potential in Lithuania and the goals that country aims to achieve by the end of the period, while setting the estimated goal for every ten years, as it was done in Germany, this way making the intentions of the government more clear for the potential investors.

The strategy has to take into account the legislation of Lithuania as well as the economic potential for supporting the solar energy sector. Planning and establishing the goal for every year or every ten years and even for the end of the planned period it has to be understood, that the more understandable and detail the strategy is, the more attractive it becomes for the investors. The strategy does not have to show unnaturally big numbers in order to attract investment. It has to be realistic. Mainly, it is important to ensure the investors that the solar energy sector has the future in Lithuania. For that the numbers and goals have to be as close to the reality as possible and the measures have to be clear. The most important is to state the aims of the strategy, aims of the sector development and the vision of the sector, assuring that the aims are as much as important for the country, the government and other relevant authorities as it is stated in the strategy and law. The investor has to be ensured that the government and other authorities are committed to fulfilling their responsibilities.

3.1.2. Improving the Feed-In Tariff System

As mentioned earlier, the feed-in tariff has appeared to be the main reason for people to invest in solar energy sector of Lithuania in past year, especially in solar energy plants with the capacity up to 30 kW. This was triggered by the establishment of the new feed-in tariffs with high prices (highest for the 30 kW installations) that led the investment to return in less than 5 years, while the support at the same price was promised for the period of 12 years.

Later, after the changes done by the new government, the tariffs were lowered suddenly and several times. The new government has also changed the conditions that were set in the law initially, which resulted to the situation that was disappointing and angering to the investors. Due to this, investors claimed to have suffered big losses and that most of them intended to sue the government and demand to cover the loss. With the current situation in the solar energy sector of Lithuania, when the tariffs are no longer considered attractive and other conditions that were initially not attractive enough for these and most of other investors, the sector needs improvement on the feed-in tariff system.

In order to increase the attractiveness of the solar energy sector for investors the feed-in tariff system could be readjusted according to the analysis of the conditions set in the countries that are considered as advanced in this field. Changes should be made using the mentioned examples, since the experience of those countries show the measures to be successful. Attempting to improve the current feed-in tariff system it is not enough to change the prices for the electricity bought from the energy producers. The changes must also be done regarding the period for which the support is granted.

According to the analysis conducted in this study, the support system for solar energy sector in Germany is considered as successful and, therefore, could be used as an example for the solar energy sector support system in Lithuania. Due to this, the two main suggestions arise: to change the period that is set for the guaranteed support in the feed-in tariff system and to establish and annually publicize the concrete amount of money designated for this support.

First of all, to be clear regarding the prices in the feed-in tariff system, increasing the prices or returning them to the initial level is not suggested in this study due to the several claims of the government stating that the high prices would have caused the need of additional 26 million Euro every year in case all the permits issued to this day would eventually have led to the same number of solar energy plants built in Lithuania.[83] On the other hand, it is hard to estimate the probability of all the permits resulting to the plants built, since many of the permits were issued to individuals or companies that were claimed as not capable of actually completing their projects and applied in order to just get a chance of receiving the highest price for the electricity.[84]

In order to increase the attractiveness of the feed-in tariff system, according to the system adopted and successfully functioning in Germany, and this way increase the attractiveness of the solar energy sector in general it is recommended to increase the period that is currently set for the feed-in tariff in Lithuania. Currently it is granted that the same feed-in tariff will be paid to the producer for 12 years. The suggestion is to increase this period to 20 years, as it is set in Germany (while even 25 years are set in the UK).

Increasing the period will allow to keep the lower prices without the disappointment from investors, since the longer period creates the conditions for the guaranteed return on investments. Recently, with the 12 years condition, the prices were set high in order to attract investment. After estimating that it will result to the lack of funds the prices were lowered to the appropriate for the funds level. With the same level it is possible to attract the investors with the longer granted period of support. Since the lowered prices now have led to the increase of the pay off period, most investors were indignant. But having increased the period of the support allows increasing the total return and this way it compensates the lowering of the prices.

For example: with the price set in 2012 of 0.42 €/kWh, the approximate price of the project of building the solar energy plant - 46 376 € and the smallest amount of energy produced per year by the 30 kW solar power plant - 27600 kWh we get the period of the investment return to be 46 376 / (27600×0.42) = **4** years. The rest of the period for 8 years the investor gets the profit. On the other hand, with the current price of 0.16 €/kWh, same price of the project, same capacity and produced amount of electricity we get that the investment will return in 46 376 / (27600×0.16) = **10.5** years.[85] With the period of the granted support being 12 years there are only 2 years left to earn

---

[83] **Jegelevicius Linas**, *Lithuanian solar investors sue government over FIT cuts*, 19th of June 2013, viewed on 19.07.2013, available online at URL:
http://www.pv-magazine.com/news/details/beitrag/lithuanian-solar-investors-sue-government-over-fit-cuts_100011774/#axzz2a0Tkwkda
[84] **Maciulis V.**, *Saulės energetika Lietuvoje: Kas Gi Nutiko? (eng. Solar Energetics in Lithuania: What Happened?)*, 22nd of February 2013, viewed on 23.07.2013, available online at URL:
http://verslas.delfi.lt/energetika/vmaciulis-saules-energetika-lietuvoje-kas-gi-nutiko.d?id=60749505
[85] **UAB Informacinių technologijų pasaulis**, *Integruotos saulės elektrinės (eng. Integrated solar energy plants)*, 2013, viewed on 30.07.2013, available online at URL: http://www.sauleselektrines.lt/lt/produktai/integruotos-saul%C4%97s-elektrin%C4%97s

the profit in comparison with 8 years in previous conditions. In order to return the previous attractiveness of investment the period should be increased. The equally attractive, considering current conditions, would be the period of at least 18.5 years, which can be rounded to be 20 years. The longer period and the guarantee of the profit will increase the security of investment and, therefore, the attractiveness of it.

Another suggestion would be to annually announce the estimated amount of money designated for the support of the sector (the payment of the feed-in tariff). In such case investors will be able to estimate the situation and it will make it easier for the government and other authorities to control the situation in the solar energy sector support system, while minimising the possibility of the occurrence of the recent negative situation once again.

3.1.3. Improving the System of Limitation

In order to make it easier to control the situation in the solar energy sector and the solar energy sector support system, it is highly important to improve the limitation system in the sector development. In the renewable energy law, as it was mentioned in the analysis, the country has set the limit on the increase of installed capacity to be 10 MW. This limit is stated to not include the installations with the capacity up to 30 kW. Due to the recent events in Lithuania the huge amount of permits were issued for the projects of 30 kW installations.[86] As a result, the total of the installed capacity increased, but without being included in the initial calculation and the limit set in the law it created the situation that led to current changes and problems.

The suggestion would be to estimate a limit for the installations with the capacity up to 30 kW as well or to change the current limit or its calculation by including all the capacities into the limit currently set in the law. This would allow avoiding the recent problems to appear in the future. Moreover, it would help to control the development of the solar energy sector and plan the future actions, while aiming to maintain or promote the development.

---

[86] **Renewable Energy Law**, 12th of May 2011, viewed on 23.07.2013, available online at URL: http://www3.lrs.lt/pls/inter3/dokpaieska.showdoc_l?p_id=398874

3.1.4. Avoiding Making Sudden Changes

According to the analysis, the changes in the solar energy sector of Lithuania were sudden and unexpected for the investors that led to the indignation among the investors and other market participants. Most of them claimed to be disappointed and no longer see the possibilities of the further development of the sector due to the recent changes.[87]

The analysis has also shown that the development in Germany, the country that was chosen as an example for Lithuania, was and continues to be gradual. There were no sudden changes in the law or the support system. The feed-in tariff is being lowered gradually by small amounts of money in order to avoid bigger changes and harm the attractiveness of investment or the sector in general.

Using Germany as an example of successful measures resulting to their solar energy sector being one of the most advanced in the world the suggestion would be to avoid making sudden changes in the solar energy sector of Lithuania.

After establishing the strategy, improving the support and limitation systems there should be a gradual development in the sector. Any sudden change might once again result to the situation that appeared in Lithuania in recent past. Constantly planning, estimating and controlling the development of the sector will help to avoid sudden changes or a decrease in the trust of the investors or in the attractiveness of the solar energy sector in general.

3.2. Evaluating the model

The actions suggested in the model are based on the analysis conducted in this study, the assessment of situation and policies in the solar energy sector of Lithuania, the experience of the most advanced solar energy sector in the European Union - solar energy sector of Germany - and the results of screening out the differences and disadvantages that helped to find the measures for the solar energy sector development in Lithuania that might be improved.

The model suggests several measures and/or actions that could be taken in order to increase the attractiveness of the solar energy sector. According to the analysis, the actions and measures suggested in the model are eligible for Lithuania and expected to have a positive influence on the development of the sector as well as its attractiveness for investors.

The model is considered to fulfil the aim of this study - creating the development model for solar energy sector of Lithuania basing it on the experience of the countries of the European Union, while choosing the best example of the most advanced sector.

---

[87] **Degutis G.** *Lietuvoje saulės energetikos plėtrai vietos neliko (eng. There is no more place left in Lithuania for the development of the solar energetics)*, 25th of January 2013, viewed on 25.07.2013, available online at URL: http://vz.lt/?PublicationId=f413933c-5fe5-4745-9db1-bb1b5bdd866e

CONCLUSIONS AND SUGGESTIONS

- Solar energy sector being a part of renewable energy sector makes it highly relevant and important in the European Union. Promoting the development of these sectors is high priority in this region. Lithuania being a member of the EU is obliged to achieve certain results in developing the solar energy sector as well as the sector of renewable energy in general.

- The relevance and importance of the solar energy is also expressed in its tendency to be more popular and have a bigger demand, since the use of this energy is wider (it can be used for bigger number of purposes than others) and bigger variety of products can be produced from it. Special features of the solar energy increase its attractiveness: minimal maintenance required, low cost, comparably small and highly modular technologies, inexhaustibility of source, possibility of producing heat and electricity, absence of noise as well as polluting gases or other harmful emissions, etc.

- Analysing and describing the advantages of solar energy to the security of the energy supply, effect to the economy, protection of the environment, describing technologies and main differences, production, use and need of the energy allowed building a better understanding of the sector and its relevance.

- Analysing the situation regarding solar energy in the European Union allowed screening out the most advanced and the most successful solar energy sectors in the region. Comparing the data collected about each of the most advanced countries regarding the relevant to the topic of this study aspects resulted to the choice of one country - Germany. It was chosen to be an example for the development of solar energy sector of Lithuania.

- Further analysis of the solar energy sector and other relevant aspects in Germany created a general view of the situation as well as an understanding of the development process in the sector, while presenting the main measures and actions that have led the sector into current success.

- Analysing the solar energy sector and other relevant aspects in Lithuania allowed understanding the general situation, familiarising with policies and measures used for sector development promotion and screening out the disadvantages and other issues of the sector that might or should be improved in order to increase the attractiveness of the sector and create a positive effect on the development of the sector in the future.

- Comparing the results of analysis of Lithuania and Germany allowed seeing more clearly what might be the issue or what could be the way of improving the situation in Lithuania.

The initial thinking that the problem in Lithuania was bad funding and feed-in tariff being too low has not been confirmed by the analysis and the comparison. The results showed the tariffs being higher than in Germany and the other aspects being the possible reason of problems.

- As a result of the analysis and comparisons the main issues were discovered and chosen to be discussed further. The development model was chosen as the way of presenting and describing the suggestions for improvement. The model is based on the number of steps or actions that are considered as recommendable in order to increase the attractiveness of the solar energy sector of Lithuania while using examples from the solar energy sector of Germany.

- The model created in the study and the suggestions that it contains are considered to be the main possible and available for Lithuania measures and actions while aiming to make positive changes in the development of the solar energy sector of Lithuania.

- The aim and all of the tasks set for this study where fulfilled during writing of this study, acknowledged and evaluated in the conclusions.

- While making general conclusions it is also important to mention the effect of the analysis conducted for this study. After reviewing and analysing a huge amount of different sources and choosing the ones that seemed more suitable and trustworthy and conducting the detailed analysis of chosen sources the general understanding of the solar energy sector and everything that is relevant to it was changed and, in a way, built from the beginning. Having created such understanding, the interest in this topic increased and remains for further wider analysis.

BIBLIOGRAPHY

1. **Curry Andrew**, *Can You Have Too Much Solar Energy?*, March of 2013, viewed on 20.07.2013, available online at URL: http://www.slate.com/articles/health_and_science/alternative_energy/2013/03/solar_power_in_germany_how_a_cloudy_country_became_the_world_leader_in_solar.html

2. **Damian Carrington**, *Wind and solar power capacity surge*, 14 February 2013, viewed on 29.06.2013, available online at URL: http://www.euractiv.com/energy/wind-power-capacity-grew-20-glob-news-517720

3. **Degutis G.** *Lietuvoje saulės energetikos plėtrai vietos neliko (eng. There is no more place left in Lithuania for the development of the solar energetics)*, 25th of January 2013, viewed on 25.07.2013, available online at URL: http://vz.lt/?PublicationId=f413933c-5fe5-4745-9db1-bb1b5bdd866e

4. **DIRECTIVE 2001/77/EC OF THE EUROPEAN PARLIAMENT AND OF THE COUNCIL of 27 September 2001 on the promotion of electricity produced from renewable energy sources in the internal electricity market**, *Article 1*, 2001, p.5, viewed on 23.06.2013, available online at URL: http://europa.eu/legislation_summaries/energy/renewable_energy/l27035_en.htm

5. **DIRECTIVE 2009/28/EC OF THE EUROPEAN PARLIAMENT AND OF THE COUNCIL of 23 April 2009 on the promotion of the use of energy from renewable sources and amending and subsequently repealing Directives 2001/77/EC and 2003/30/EC**, *Article 1*, 2009, p.11, viewed on 23.06.2013, available online at URL: http://eur-lex.europa.eu/LexUriServ/LexUriServ.do?uri=OJ:L:2009:140:0016:0062:EN:PDF

6. **Energy Charter Secretariat**, *Investment Climate and Market Structure Review in the Energy Sector of Lithuania*, 2013, p.11, viewed on 19.07.2013, available online at URL: http://www.encharter.org/fileadmin/user_upload/Publications/Lithuania_ICMS_2013_ENG.pdf

7. **Epsea.org**, *Germany Invest on Alternative Energy: Leading to Make the World Greener*, 2011, viewed on 18.07.2013, available online at URL: http://www.epsea.org/germany-invest-on-alternative-energy-leading-to-make-the-world-greener.html

8. **Ernst & Young Global Limited**, *Renewable energy country attractiveness index: May 2013*, Issue 37, 2013, p.11, p.16, p.38, viewed on 12.06.2013, available online at URL: http://www.ey.com/Publication/vwLUAssets/Renewable_energy_country_attractiveness_indices_-_Issue_37/$FILE/RECAI-May-2013.pdf

9. **Europa.eu** *"Renewable energy: the promotion of electricity from renewable energy sources"*, 20.01.2011, viewed on 20.06.2013, available online at URL: http://europa.eu/legislation_summaries/energy/renewable_energy/l27035_en.htm

10. **Europa.eu**, *"Renewable energy: Promotion of the use of energy from renewable sources"*, 09.07.2010, viewed on 22.06.2013, available online at URL: http://europa.eu/legislation_summaries/energy/renewable_energy/en0009_en.htm

11. **European Commission**, *Quarterly Report on European Electricity Markets*, March 2013, p.3, viewed on 16.06.2013, available online at URL: http://ec.europa.eu/energy/observatory/electricity/doc/20130611_q1_quarterly_report_on_european_electricity_markets.pdf

12. **European Commission**, *Renewables Make the Difference,* 2011, p.14, p.23, viewed on 23.06.2013, available online at URL: http://www.energy.eu/publications/Renewables-make-the-difference-2011.pdf

13. **European Photovoltaic Industry Association**, *Global Market Outlook For Photovoltaics Until 2016,* 2012, viewed on 10.06.2013, available online at URL: http://www.helapco.gr/ims/file/reports/Global%20Market%20Outlook%202016.pdf

14. **Federal Ministry for the Environment,** *Nature Conservation and Nuclear Safety, Renewable Energy Sources Act (EEG) 2012,* May. 2013, viewed on 26.07.2013, available online at URL: http://www.erneuerbare-energien.de/en/unser-service/mediathek/downloads/detailview/artikel/renewable-energy-sources-act-eeg-2012/

15. **German Solar Industry Association,** *Policy framework: The Renewable Energy Sources Act (EEG),* 2013, viewed on 18.07.2013, available online at URL: http://www.solarwirtschaft.de/en/photovoltaic-market/political-framework.html

16. **German Solar Industry Association,** *Statistic data on the German solar power (photovoltaic) industry,* July 2013, p.2, p.3, viewed on 20.07.2013, available online at URL: http://www.solarwirtschaft.de/fileadmin/media/pdf/2013_2_BSW-Solar_fact_sheet_solar_power.pdf

17. **Germany Trade & Invest**, *Industry Overview: The Photovoltaic Market in Germany,* Issue 2013/2014, p.8, viewed on 24.07.2013, available online at URL: http://www.gtai.de/GTAI/Content/EN/Invest/_SharedDocs/Downloads/GTAI/Industry-overviews/the-photovoltaic-market-in-germany.pdf

18. **International Energy Agency**, *IEA says further action is needed if Germany's Energiewende is to maintain a balance between sustainability, affordability and competitiveness*, 24 May 2013, viewed on 29.07.2013, available online at URL: http://www.iea.org/newsroomandevents/pressreleases/2013/may/name,38340,en.html

19. **International Renewable Energy Agency**, *Renewable Energy Technologies: Cost Analysis Series. Solar Photovoltaics,* 2012, viewed on 15.06.2013, available online at URL: http://www.irena.org/DocumentDownloads/Publications/RE_Technologies_Cost_Analysis-SOLAR_PV.pdf

20. **Jegelevicius Linas**, *Lithuanian solar investors sue government over FIT cuts*, 19th of June 2013, viewed on 19.07.2013, available online at URL: http://www.pv-magazine.com/news/details/beitrag/lithuanian-solar-investors-sue-government-over-fit-cuts_100011774/#axzz2a0Tkwlda

21. **KPMG International Cooperative**, *Green power 2012: The KPMG renewable energy M&A report*, 2012, p.11, p.13, p.29, viewed on 25.06.2013, available online at URL: http://www.kpmg.com/CZ/cs/IssuesAndInsights/ArticlesPublications/Press-releases/Documents/KPMG-Green-Power-2012.pdf

22. **Lehmann Paul / Creutzig Felix / Ehlers Melf-Hinrich / Friedrichsen Nele / Heuson Clemens / Hirth Lion / Pietzcker Robert** "Carbon Lock-Out: Advancing Renewable Energy Policy in Europe", 15.02.2012, p.324

23. **Maciulis V.**, *Saulės energetika Lietuvoje: Kas Gi Nutiko? (eng. Solar Energetics in Lithuania: What Happened?)*, 22nd of February 2013, viewed on 23.07.2013, available online at URL: http://verslas.delfi.lt/energetika/vmaciulis-saules-energetika-lietuvoje-kas-gi-nutiko.d?id=60749505

24. **Marquardt Jens**, *Renewable Energy Projects - The German Energy Transition in a Multi-Level Perspective*, 2012, viewed on 21.07.2013, available online at URL: http://www.academia.edu/2276998/Renewable_Energy_Projects_-_The_German_Energy_Transition_in_a_Multi-Level_Perspective

25. **National Control Commission for Prices and Energy**, *Commission approval of tariffs for electricity produced from renewable energy resources for 2013 3rd quarter*, 2013, viewed on 11.07.2013, available online at URL: http://www.regula.lt/lt/naujienos/index.php?full=yes&id=49624

26. **National Control Commission for Prices and Energy**, *Commission approval of tariffs for electricity produced from renewable energy resources*, 2012, viewed on 11.07.2013, available online at URL: http://www.regula.lt/lt/naujienos/index.php?full=yes&id=14048

27. **Poderis Justinas,** *Investment Opportunities in Renewable Energy Resources in Lithuania*, 5th of October 2012, viewed on 05.07.2013, available online at URL: http://www.terralex.org/publication/p459bcb8ea3/investment-opportunities-in-renewable-energy-resources-in-lithuania

28. **Renewable Energy Law,** 12th of May 2011, viewed on 23.07.2013, available online at URL: http://www3.lrs.lt/pls/inter3/dokpaieska.showdoc_l?p_id=398874

29. **Report for the European Commission,** *Renewable energy progress and biofuels sustainability*, September 2012, p.113, p.116, viewed on 07.07.2013, available online at URL: http://ec.europa.eu/energy/renewables/reports/doc/2013_renewable_energy_progress.pdf

30. **Sapronaitytė Inga ,** *Opinion: Will renewable energy sources lead towards energy security?*, 12th June 2013, viewed on 08.07.2013, available online at URL: http://www.lithuaniatribune.com/41337/opinion-will-renewable-energy-sources-lead-towards-energy-security-201341337/

31. **Schlütersche Verlagsgesellschaft mbH & Co. KG,** *German PV drops to 15 cents max*, 2nd of May 2013, viewed on 22.07.2013, available online at URL: http://www.renewablesinternational.net/german-pv-drops-to-15-cents-max/150/510/62457/

32. **The World Bank,** *Ease of Doing Business in Germany*, 2013, viewed on 27.07.2013, available online at URL: http://www.doingbusiness.org/data/exploreeconomies/germany

33. **The World Bank,** *Economy Profile: Lithuania, 10th Edition*, 2013, p.7, viewed on 27.07.2013, available online at URL: http://www.doingbusiness.org/~/media/giawb/doing%20business/documents/profiles/country/LTU.pdf

34. **UAB Informacinių technologijų pasaulis,** *Integruotos saulės elektrinės (eng. Integrated solar energy plants)*, 2013, viewed on 30.07.2013, available online at URL: http://www.sauleselektrines.lt/lt/produktai/integruotos-saul%C4%97s-elektrin%C4%97s

35. **Winkel Thomas, Rathmann Max, Ragwitz Mario, Steinhilber Simone, Winkler Jenny, Resch Gustav, Panzer Christian, Busch Sebastian, Konstantinaviciute Inga,** *Renewable Energy Policy Country Profiles*, 2012, viewed on 18.06.2013, available online at URL: http://www.ecofys.com/files/files/ecofys_re-shaping_country_profiles_2011.pdf